AVIATION INVESTMENT

To my father

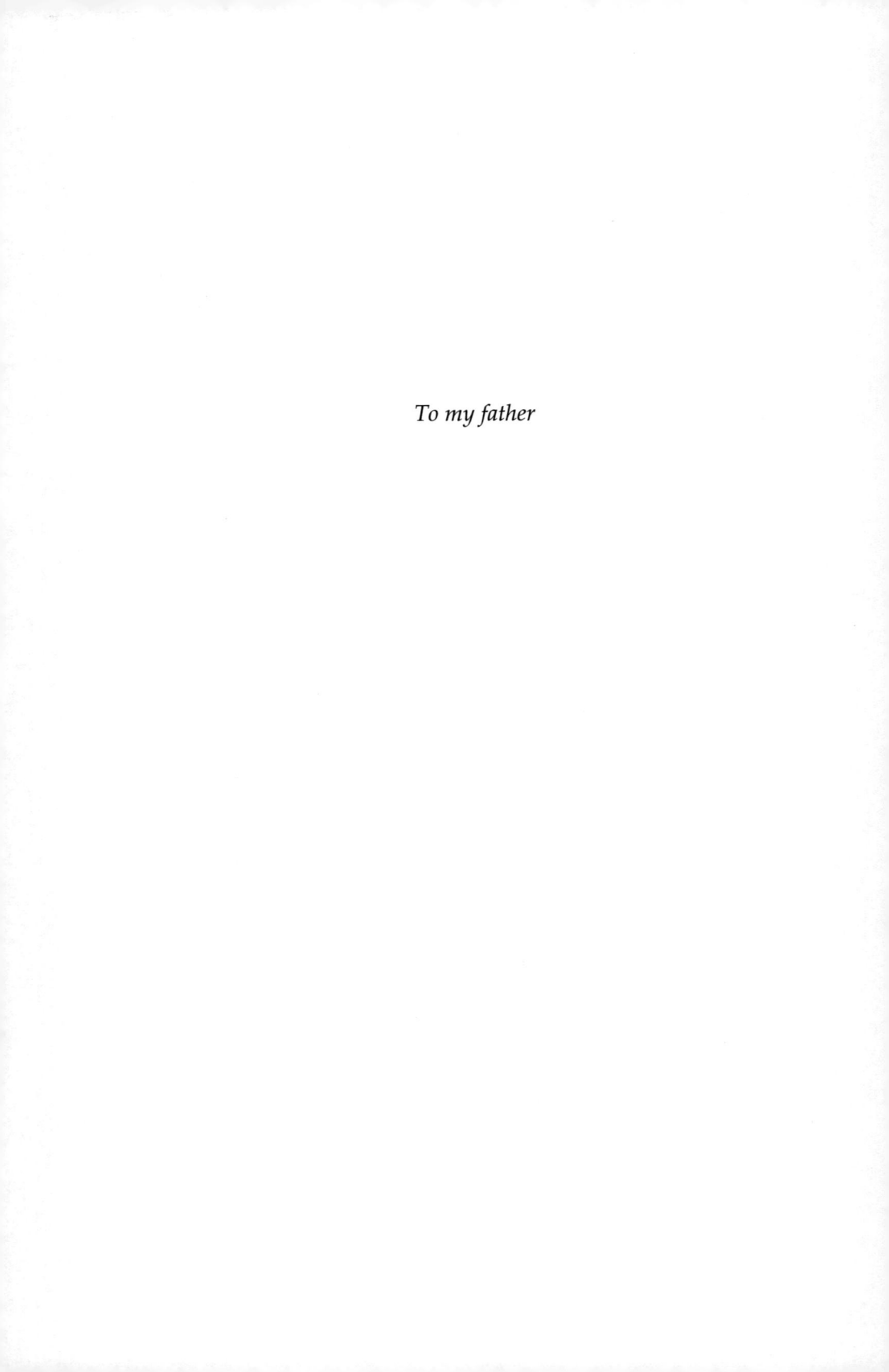

Aviation Investment
Economic Appraisal for Airports, Air Traffic Management, Airlines and Aeronautics

DORAMAS JORGE-CALDERÓN
European Investment Bank

ASHGATE

© Doramas Jorge-Calderón 2014

All rights reserved. No part of this publication may be reproduced, stored in a retrieval system or transmitted in any form or by any means, electronic, mechanical, photocopying, recording or otherwise without the prior permission of the publisher.

Doramas Jorge-Calderón has asserted his right under the Copyright, Designs and Patents Act, 1988, to be identified as the author of this work.

Published by
Ashgate Publishing Limited
Wey Court East
Union Road
Farnham
Surrey, GU9 7PT
England

Ashgate Publishing Company
110 Cherry Street
Suite 3-1
Burlington, VT 05401-3818
USA

www.ashgate.com

British Library Cataloguing in Publication Data
A catalogue record for this book is available from the British Library.

The Library of Congress has cataloged the printed edition as follows:
Library of Congress data has been applied for.

ISBN 9781472421302 (hbk)
ISBN 9781472421319 (ebk – PDF)
ISBN 9781472421326 (ebk – ePUB)

Printed in the United Kingdom by Henry Ling Limited, at the Dorset Press, Dorchester, DT1 1HD

Contents

List of Figures		*vii*
List of Tables		*ix*
Preface		*xi*
List of Abbreviations		*xix*

1	**Introduction**	1
	1 Reasons to Invest in Aviation	1
	2 Financial and Economic Returns	4
	3 Discount Rate, Risk and Uncertainty	10
	4 Additional Considerations	12

2	**Identifying Benefits**	15
	1 Air Transport as an Intermediate Service	15
	2 Travel Time	16
	3 The Money Cost of Travel	19
	4 Accident Risk	22
	5 Externalities	23
	6 The Generalised Cost of Transport	26
	7 Wider Economic Benefits	29

3	**The Basic Framework**	41
	Introduction	41
	1 Landside Investments	41
	2 Airside Investments	47
	3 Scenario-building	50

4	**Airports**	69
	Introduction	69
	1 A Greenfield Airport	70
	2 Involving the Private Sector (1): No Room for Capital Investment	84
	3 Terminal Capacity Expansion	92
	4 Involving the Private Sector (2): Room for Capital Investment	100
	5 The Incentive to Overinvest	101
	6 Enlarging a Runway	115

		7 Adding a Runway	126
		8 Involving the Private Sector (3): Regulatory versus Competitive Outcome	138
5	**Air Traffic Management**		**141**
	Introduction		141
	1 Greater Movement Capacity		141
	2 Involving the Private Sector (4): Pricing Policy		149
	3 Flight Efficiency		154
6	**Airlines**		**159**
	Introduction		159
	1 Fleet Replacement		162
	2 Fleet Expansion		167
	3 The Value of Air Transport		177
7	**Aeronautics**		**191**
	Introduction		191
	1 Low Uncertainty		194
	2 High Uncertainty		209
8	**Concluding Remarks**		**223**

References 229
Index 233

List of Figures

2.1	Effect of a tax on the money cost and economic cost of an input	21
2.2	Effects of a project on secondary markets	34
3.1	Demand and supply in landside capacity provision	42
3.2	Costs in airside capacity provision	49
3.3	Alternative counterfactual capacity conditions in the presence of market power	61
4.1	Effects of increasing airport aeronautical charges on user and airport surpluses	90
4.2	Effect of an increase in income on the investment case for a new runway	138
6.1	Option price and value at expiration and investment returns	172
7.1	Effect of introducing competition in a monopolised aircraft market	200
7.2	Treatment of environmental costs in the economic appraisal of investments in competitive aircraft manufacturing markets	208
7.3	Binomial tree for the financial real option value of an aircraft engine project	214
7.4	Binomial tree for the economic real option value of an aircraft engine project	222

List of Tables

1.1	Financial and economic calculations of return on investment: differences and linkages	9
2.1	Components of generalised cost of transport as used in this book	29
3.1	Competitive conditions and scenario-building in investment appraisal	55
4.1	Estimating traffic-generating potential of town A	70
4.2	Estimating deterred (or generated) traffic in town A	72
4.3	Financial and economic returns of a greenfield airport project	76
4.4	Financial and economic returns of the greenfield airport project with higher aeronautical charges	86
4.5	Financial and economic returns of the terminal capacity expansion project	94
4.6	Financial and economic returns of terminal expansion with larger capacity but with no price increase	104
4.7	Financial and economic returns of terminal expansion with larger capacity and with price increase	111
4.8	Financial and economic returns of a runway enlargement project with no change in aeronautical charges	118
4.9	Returns on a runway enlargement project with change in aeronautical charges and limited airline competition	123
4.10	Returns on a runway enlargement project with change in aeronautical charges and competitive airline market	124
4.11	Economic returns of adding a new runway	134
4.12	Economic returns of adding a new runway with faster growth in aircraft size without the project	136
5.1	Economic returns on an ATM project aimed at increasing aircraft movement capacity	144
5.2	Financial returns for an ANSP of investing in greater aircraft movement capacity	151
5.3	Economic returns on an ATM investment project aimed at improving flight efficiency	156
6.1	Financial and economic returns on a fleet replacement project with external emissions costs	165
6.2	Financial and economic returns on a fleet replacement project with emissions costs internalised	166

6.3	Traffic and investment return scenarios for an airline considering an expansion of its fleet	169
6.4	Contribution to GDP and generalised cost of a hypothetical passenger trip across various route lengths	182
6.5	Returns on investing in an air service	186
7.1	Returns on investment in the aeronautical sector under alternative scenarios regarding competition and external costs	198

Preface

Aviation has significant advantages over alternative transport modes. The ability to offer fast, reliable services, largely independently of geographical obstacles, means that the degree to which it improves accessibility worldwide cannot be matched by other transport technologies, so that aviation has become a very distinctive source of value. As a result, society and the economy at large benefit through the widening of the scope of markets, fostering the generation of wealth and the enriching of lifestyles. Today, aviation is a necessary component of commercial and cultural activity. Without it, the functioning of modern societies and economies as we know them would be fundamentally altered.

Because of the strong competitive advantage as a sector, particularly for long-distance passenger travel, aviation investments can be very profitable. Whereas some sub-sectors within aviation, notably airlines, have a mixed reputation with private investors, the underlying competitive advantage of aviation means that there are pockets of strong pricing power. Economic regulation to cap prices is frequent, as is, increasingly, discretionary taxation to raise revenues for governments. Moreover, whereas it is already a very large industry, aviation is expected to continue growing, probably doubling in size over the next 15 to 20 years. Much is made, and rightly so, about aviation not paying for its full environmental cost. Still, the distinctiveness of the aviation product is such that making aviation pay fully for its environmental cost would only marginally affect its viability.

Despite strong value generation, competitive advantage and high growth rates, substantial amounts of resources can also be wasted in unviable investments in both the private and public sectors. Aviation is a capital-intensive sector, where investments can involve large sums of money and where debt tends to account for an important share of the financial structure of industry operators. High financial gearing underscores the need for sound investment decisions, as both gains and losses are leveraged. Bad, large, leveraged investments can be instrumental in putting private companies or local economies in severe financial hardship. Beyond financial considerations, bad investments represent a waste of resources that society could deploy in other, more productive activities. Conversely, good, large, leveraged investments can help generate substantial profits and transform local economies for the better.

Private sector investments are generally appraised through standard business plans, including estimates of the financial return and present value of the investment, a financing plan and a risk assessment. However, transport operations are characterised by taxes, subsidies and externalities – both positive and negative. Also, situations of monopolistic competition, when combined with price regulation, can result in substantial non-monetised user benefits. As a result, financial analysis cannot be expected to measure the total private value generated by investments, or their viability for the economy at large. Financially profitable operations may reflect financial transfers among stakeholders and protection to certain operators rather than genuine value generation. Conversely, financial losses can mask projects that are still worth carrying out because of the non-monetised value they generate. Therefore, governments are unlikely to rely on financial appraisals alone, often requiring cost-benefit analyses – also called economic appraisals – in order to evaluate the underlying economic viability of investments for society at large.

All too often, the private sector is only interested in cost-benefit analysis as a marketing tool to help make a case to the government for a project, generally involving government financial support or protection from competition. Otherwise it deems economic appraisal a largely academic exercise with little or no relevance for the business case of the project. This reflects the general misunderstanding among transport managers of what an economic appraisal conveys. At times, managers putting forward the results of economic appraisals emphasise elements such as the jobs created by the project, the expenditure by tourists and other benefits to the wider economy, and savings on environmental emissions.

In reality, even when any such alleged benefits are legitimate (e.g. jobs are a cost, not a benefit; and expenditure by tourists is not a benefit) they tend to be a small proportion of project benefits. The actual benefits to both the private sector and society at large tend to be much more immediate and relevant to the operator. By removing distortions, taking into account the most immediate externalities (in practice most other alleged externalities tend not to be legitimate) and measuring non-monetised user benefits, an economic appraisal unveils the value of the sustainable competitive advantage of an operation, its pricing power and the risks that may hide behind market distortions. In addition, the mechanics of calculating the full economic returns of a project informs the demand forecasts used in the financial analysis. For the private sector investment analyst, financial and economic analyses complement each other. Financial appraisal constitutes the building block from which to start building the economic appraisal. In turn, the economic appraisal gives a comprehensive picture of the intrinsic viability of the investment, yielding important information to the private investor regarding profit potential and sources of risk. For the public sector investment analyst, the economic appraisal is the central test on which to base the investment recommendation.

This book combines standard methods of financial, cost-benefit (i.e. economic), and real-option analyses, and applies them to the appraisal of the financial and economic viability of aviation investments. Also, it highlights the relevance of economic analysis to private-sector financial appraisals, applying cost-benefit analysis to sectors where it is used more rarely, including airlines and aeronautics.

The term 'investment' is used in its economic sense: that is, the assignment of resources to produce capital, where capital is any asset, physical or not, used to produce useful goods or services. The book deals primarily with physical capital assets, including airports, air traffic management (ATM) infrastructure, aircraft and aircraft manufacturing plants, but also with intellectual capital, including research into aircraft technology. In so doing, it deals with an array of different conditions regarding technology and competition, as follows:

- Airports: infrastructure operations with a substantial component of sunk costs, operating as monopolies or under monopolistic competition, but which are becoming increasingly competitive;
- ATM: infrastructure operations often in a position of natural monopoly;
- Airlines: a capital-intensive industry in the service sector, with low barriers to entry and exit and limited scope for product differentiation, which make its markets very competitive;
- Aerospace: a high-tech manufacturing sector with heavy up-front investments in product development, operating under oligopolistic or monopolistic competition.

The book is aimed at public and private sector analysts concerned with appraising the financial and economic case for aviation investments, as well as to students of air transport and of applied investment appraisal. It assumes at least some training in economics. It is written using the easiest possible language but takes for granted basic knowledge of standard financial appraisal techniques and provides only short explanations of general economic appraisal topics, which are well documented in the literature.

Similarly, the book also illustrates the use of real option analysis, always including a step-by-step calculation process, but leaving justification for real options to other sources. Whereas real options are geared towards conditions of uncertainty, the book does not deal with risk analysis or management, as there is no feature in aviation that raises sector-specific issues. The techniques used in other sectors apply to aviation and the reader is referred to a well-supplied market of project risk management references.

Air transport demand forecasting and cost estimation are not discussed as they are fully covered in the applied transport economics and transport planning literature, so that to discuss then here would simply duplicate material available elsewhere. Instead, the focus is on methods of measuring investment returns, which the literature covers extensively for other modes of transport, particularly land transport modes, but less so for aviation.

The book is structured in two broad parts. The first runs from Chapter 1 to Chapter 3 and includes the conceptual framework (or theoretical background) that underpins the measurement of returns from the financial and the economic points of view. The introductory Chapter 1 provides a brief overview of the difference between financial and economic profitability, highlighting the links between them. Chapter 2 identifies the benefits of aviation projects, which fall into three groups: first, the drivers of customer value, which determine consumer surplus in the economic appraisal, which in turn underpins competitive advantage, on which any financial profitability must rest; second, external effects, which are also an important driver of economic returns and can be interpreted as signals of regulatory risk in the financial analysis; and third, the wider economic benefits of investments, a fertile source of invalid reasons to justify bad investments. The first part of the book concludes with Chapter 3, which introduces the basic theoretical framework underpinning the benefit measures.

The second part of the book consists of four chapters that address each of the aviation sectors in the subtitle of the book. Chapter 4 addresses airports, including investments to accommodate passengers and those aimed at accommodating aircraft. Chapter 5 addresses air traffic management, including investments aimed at expanding the airspace aircraft movement capacity and those aimed at improving flight efficiency. These two chapters cover, therefore, the basic infrastructure sectors of air transport. Infrastructure operations tend to be government owned; when privatised, they are normally subject to economic regulation. Included in these two chapters are four sections addressing investment issues that arise with private sector involvement (labelled 'involving the private sector'). The issues they cover include identifying when there is room for private sector participation in the investment process and when there is only justification for a management contract; differences that regulation may make to the outcome of the investment; and the relevance of pricing policy in affecting incentives and outcomes. In addition, Chapter 4, section 5, addresses the incentives to overinvest in infrastructure, which is left out of the list of private sector issues because it may also apply to the public sector.

Chapter 6 addresses airlines, usually the most competitive of the aviation sectors. Because of the extent of competition, the role for economic analysis is more limited. However, it still plays a role, particularly when addressing inter-modal competition and when estimating the value of air transport to society. In addition, options on aircraft tend to be present in airline aircraft purchase programmes, mostly due to the high degree of uncertainty regarding future market and competitive conditions. The chapter discusses the circumstances under which options are valuable, and how to value them. Chapter 7 addresses the aeronautical sector, also a competitive industry, but generally imperfectly so due to high entry barriers and sunk costs. The imperfectly competitive nature of the sector calls for special considerations in the appraisal process. Also, as with airlines, uncertainty gains relevance again, in this case via the prospects of technological innovation. Here the uncertainty and risks involved are usually a motivation for government intervention in financing such investments, which again call for the tools of economic appraisal. Finally Chapter 8 offers some concluding remarks regarding possible additions to the various appraisal methods discussed.

A book that applies welfare economics tools to practical decisions concerning the building of physical assets and the launching of programmes naturally benefits from influences from a range of fields. I am deeply grateful to Professors Ginés de Rus and Per-Olov Johansson, who kindly read an early draft. Their comments contributed to the improvement of the economics underpinning the book. Thanks also to Stephen O'Driscoll, the current European Investment Bank (EIB) in-house airports engineer, for comments on technological issues in the chapter on airports. Likewise, I should mention other colleagues at the Projects Department of the EIB with whom I have appraised aviation investment projects over the years, including in particular Klaus Heege, Alan Lynch, and Bernard Pels. I owe to them an appreciation of project conception and planning through the vantage point of aeronautical, civil and systems engineering. I would also like to thank the many professionals in promoters of aviation projects financed by the EIB with whom I have worked. I have learnt from them many of the practicalities of conceiving and implementing capital investment programmes across the various industries that constitute civil aviation.

In the book I touch upon a number of regulatory and competition issues. My exposition doubtlessly benefited from what I learned during my time at National Economic Research Associates, particularly from Ian Jones (now a UK Competition Commissioner). Thanks also to a number of anonymous referees. Many of their views are reflected in the final product.

I am also grateful to Guy Loft, my commissioning editor, and to Emily Ruskell, from Ashgate, for managing the publishing process so brilliantly, as well as to Helen Varley for her most valuable editing input.

Any remaining errors or omissions are mine. So are the opinions expressed in the book, which do not necessarily reflect those of the EIB or any other institution.

Doramas Jorge-Calderón
Luxembourg

List of Abbreviations

ALRMC	Airlines' Long-Run Marginal Cost
AMEC	Airline Marginal Environmental Cost
ANSP	Air Navigation Service Provider
AS	Aircraft Size
ASK	Available Seat-Kilometre
ATM	Air Traffic Management
CAGR	Cumulative (or Compounded) Annual (or Average) Growth Rate
CBA	Cost Benefit Analysis
CCD	Continuous Climb Departures
CDA	Continuous Descend Approaches
DCF	Discounted Cash Flow
ERR	Economic (internal) Rate of Return
EU	European Union
FAA	Federal Aviation Administration (of the United States of America)
FRR	Financial (internal) Rate of Return
GC	Generalised Cost
GDP	Gross Domestic Product
GHG	Greenhouse Gas
HSR	High Speed Rail
IATA	International Air Transport Association
ICAO	International Civil Aviation Organisation
IRR	Internal Rate of Return
LRMC	Long-Run Marginal Cost
MEC	Marginal Environmental Cost
MR	Marginal Revenues
MTOW	Maximum Take-Off Weight
NPV	Net Present Value
OPD	Optimised Profile Descent
Pax	Passengers
PV	Present Value
R&D	Research & Development
RDI	Research, Development and Innovation
ROA	Real Option(s) Analysis
ROV	Real Option Value
RPK	Revenue Passenger-Kilometre
VoT	Value of Time
WACC	Weighted Average Cost of Capital

Chapter 1
Introduction

1 Reasons to Invest in Aviation

There are three main reasons to invest in aviation and these are common to all modes of transport. They are:

1. Reducing the time it takes to transport a person or freight from one place to another (including time-saving by reducing congestion and increasing on-time departure).
2. Reducing the cost, in terms of resources used, of moving a person or freight from one place to another.
3. Improving the safety of a journey by reducing the risks inherent in physical transportation.

Comfort and quality of service are additional sources of value in transport, but are rarely in themselves a reason to invest. Instead, they tend to accompany some combination of the three main reasons.

Private sector operators develop their competitive strategies by focusing primarily on the first two reasons, and value the returns on their investment through a *financial* appraisal. The third reason is mostly relevant for promoters in countries with very poor transport conditions. Public sector investors also base their investment decisions on the very same criteria, although they widen the scope of benefits and costs beyond monetised private flows to include non-monetised private flows, as well as flows to third parties including, ultimately, society at large. Such an exercise constitutes an *economic* or socio-economic investment appraisal.

The private and public perspectives on investment – the financial and the economic, respectively – are mutually complementary in two respects. First, private financial benefits and costs offer a first approximation to economic benefits and costs. This constitutes a partial look at the flows associated with a project. Second, the economic benefits of an investment offer the private sector investor clues about untapped sources of revenue; and economic costs signal potential risks arising from market distortions and badly defined property rights. These issues are explored in section 2 of this chapter.

However, the distinction between financial and economic returns is often saddled with confusion, opening the doors to abuse. For example, the projected positive financial profitability of an investment may be touted as proof of the soundness of a project. However, what is advertised as a financially viable investment may in fact not reflect social value or a competitive advantage at all, but rather transfers from other stakeholders. After all, operators and investors may try to influence public policies in order to protect their competitive positions by erecting barriers to competition and, more generally, distorting markets, in extreme cases turning a financially non-viable project into a viable one. In such situations an economic appraisal would show that the proposed investment would be wasteful, despite the positive financial return. A second example is when politicians, for electoral reasons, may want to justify devoting public money to financially loss-making investments with arguments about all sorts of wider benefits to the local economy. On closer examination, a proper economic appraisal may show that many of the alleged wider economic benefits are invalid.

Besides the three fundamental reasons to invest in transport – including time and cost savings and safety improvements, as mentioned above – investment appraisal analysts are continuously confronted with myriad other reasons put forward to justify investments. Some of these reasons are ultimately invalid, but come mixed with elements of the three valid reasons set out above, making it hard to distil the extent to which an investment creates value, and the extent to which it constitutes waste and abuse. Arguments put forward may include the following:

- *This investment will open up our region and lead to new economic activity and industry.* This is a valid rationale insofar as it is

reflected in the three fundamental reasons. Unfortunately, it tends to open the gates to all sorts of claims to benefits that are in fact mostly invalid.
- *This is the latest technology.* The fact that a project introduces the latest technology does not make it necessarily a good investment. There may be a case for keeping the technology alive, but that does not imply its deployment.
- *And this technology will improve safety.* In aviation, the safety argument has been used over the years all too often as an excuse to preserve market power (with the accompanying economic rents) and to justify transfers. Safety does not justify any expenditure, regardless of the cost. Expenditure on safety has to be set against the value of the expected safety improvement, and investments argued for on safety grounds in circumstances where operations already meet international safety standards tend to have other motivations.
- *It will create jobs and the multiplier effect will generate more economic activity in the area.* Many of the jobs 'created' may be crowded out from other activities. Moreover, loss-making investments also 'create' jobs and unleash multiplier effects. Contrary to frequent popular discourse, jobs and multipliers are not in themselves a sound reason to invest.
- *We will bring more tourists.* Whether this is a good reason or not will depend on the cost of bringing those tourists, and the added benefits the tourists generate.
- *We need to increase market share.* Many businesses have gone bust making wasteful investments in their chase for market share rather than profit.

There are also more clearly invalid reasons for investing that are easier to spot in advance:

- *We must operate that route because an airline like ours has to be seen flying that route.* Such routes are usually found on the route maps of nationalised airlines.
- *Our neighbours have it, so we must have it.* Very often politicians will push to supply locally what a nearby region or city already has, independently of whether there is a case to

have it in the neighbouring location but not in the proposing politician's constituency (or, indeed, in neither of them).
- *Visitors must be impressed when they arrive in our country.* The funds used to impress the visitors come at the expense of other items that society may demand more urgently.

And even:

- *Passengers get the feeling of an amusement park attraction when they see this project.* It may well be that the promoter is subject to rate of return regulation, and the motive of the project at hand is to inflate the regulatory asset base of the promoter. In such cases, financing the project with debt can boost the return on equity of the promoter.

To conclude, sound financial returns and arguments with popular appeal are no guarantee that the investment will be worthwhile. The ultimate case is based on saving time, reducing costs and improving safety in ways that ensure that the benefits outweigh the costs. A project with a positive financial return and a negative economic return is likely to be fully dependent on political patronage.

2   Financial and Economic Returns

The financial appraisal of an investment project involves estimating revenues and costs, including financing costs. Such an estimate constitutes the backbone of any standard business plan. In this regard, there is nothing exceptional in the mechanics of conducting the financial appraisal of an investment in the aviation sector, or in transport in general, relative to a project in any other sector. To simplify, the financial appraisal as presented in this book ignores considerations regarding the capital structure of a project. The focus is on whether the financial resources invested in a project as a whole generate a sufficient cash return to the promoter. Projects can be thought of as being 100 per cent financed with equity capital.

Under very specific circumstances the financial return of a project also measures the economic return. When markets are competitive, are free from distortions such as taxes, subsidies or price regulations, when there are close substitutes for all goods, when an investment project is too small relative to the size of the economy to significantly alter prices, and property rights are well defined, prices reflect the benefits of an additional unit of output produced and costs reflect the resource cost of producing that unit. Private sector investors, in following expected revenues and costs in making investment decisions, will make investments that are in line with maximising not only private profit but also social welfare. That is, the investor will inadvertently be part of the proverbial 'invisible hand' whereby the pursuit of private interest leads to an allocation of resources that is socially desirable.

In such circumstances, the financial appraisal of a private sector investment analyst would be sufficient to decide whether the investment should be made from the point of view of society at large, without any need for a public sector economist to carry out any other viability test. However, in reality, prices are often distorted, substitutes may be imperfect, giving certain operators a degree of market power, and property rights are not always well defined. These issues are addressed in turn in the following paragraphs.

Prices may not reflect full resource cost because of the presence of taxes, subsidies, or regulations such as minimum wages or price caps in markets for inputs or outputs. A tax on an input, for example, means that the promoter will pay for the resource cost (the opportunity cost) of the input, plus a transfer (the tax) to the government. The price the promoter pays for the input overestimates the cost of the input to society, and therefore, as far as society is concerned, this price cannot be taken as the basis for making a sound allocation of scarce resources since the taxed input would tend to be consumed less than would be socially desirable. A subsidy on an input would have the opposite effect. Similarly, price regulation, such as price ceilings or floors, may imply that the price does not reflect the scarcity of the input. Prices may instead reflect a market outcome that over- or under-supplies the good.

When property rights are not well defined, a market transaction involving a buyer and a seller may interfere with the rights of a third party that does not voluntarily take part in the transaction. These impacts to third parties are called 'externalities', in the sense that they are external to the parties that voluntarily agree a transaction. In the case of aviation the main examples of potential externalities concern the environment, including emissions of greenhouse gases, air-polluting particles and noise. When the property rights of third parties are well defined, the parties involved in the transaction will also have to pay, via taxes or direct compensation, to the third parties affected by the transaction.

It should be noted that effects on third parties may not only be negative. Projects can have positive external effects, such as knowledge spillover effects from investments in aviation research and development (R&D). There can also be beneficiary price effects, as when a project is large enough to affect the price of one of its inputs in the presence of cost economies in the production of that input. The higher demand for of the input brought about by the project would lower the price of the input, yielding productivity gains to other firms, which are unrelated to the project but also use that input.

Finally, when the products supplied by competitors are not close substitutes, consumers can experience a cost in switching from one producer to another. In such situations, if supply is lumpy (i.e. there are indivisibilities) competitive markets may not work well to address supply shortages, giving incumbents an element of pricing power that can be abused. An example may be an airport (supply is lumpy: capacity cannot be doubled at short notice) that is a monopolist in a city, and users have as an alternative another airport two hours' drive away (constituting a switching cost). The monopolist airport could adjust prices in order to try to convert all of the cost of switching into extra revenues (extracting rents through market power).

In recognition of this monopoly power, the prices offered by the airport (aeronautical charges) are often regulated by the government, leaving such switching costs un-monetised. The switching costs represent a resource use (time to drive to the alternative airport and operating cost of the vehicle to reach that other airport), so much so that the airport user would be willing to pay in order to avoid it. Such willingness to pay, however, remains unregistered by the revenues or costs of the project, and therefore ignored in a financial appraisal. Whereas they are un-monetised, the switching costs measure consumer willingness to pay – over and above existing prices (aeronautical charges) – to continue using the airport, before switching to a competing service. Switching costs constitute, therefore, a measure of the competitive advantage of the airport, that is, how much customers value the distinctive characteristics of the service offered by the operator (in this case consisting largely of location, or proximity) over and above those of its competitors and, therefore, how much extra they would be willing to pay to the airport before switching to the next best competitor.

An economic appraisal aims at quantifying the three distortions mentioned above, and incorporates them in the calculation of project returns. It attempts to work through price distortions, inefficient property rights and unobserved willingness to pay, in order to register the actual resources used by the project and the actual benefits produced by it. It is, in other words, an attempt to estimate the net benefit of the investment to society (where value to society is largely reflected by the use of the facility) when the presence of market imperfections leaves the estimate of financial return incapable of answering that question. Fortunately, the tools of economic appraisal are very apt for application to transport projects, including aviation. The standard technique for economic appraisal is cost-benefit analysis (CBA).

The literature on CBA is well developed, often extending to application to transport projects.[1] Table 1.1 below summarises the main differences between financial and economic appraisals. While it is merely a summary table, it gives a flavour of where to pay attention in order to avoid frequent sources of confusion in the calculation process. Chapters 2 and 3 discuss some of the issues that merit special attention in the context of aviation projects. Section 3 of this chapter deals with the discount rate

---

1 See Boardman et al. 2014, de Rus 2010 and Campbell and Brown 2003.

and the related subject of risk and uncertainty. These are topics that are not particular to aviation, and this introductory chapter merely outlines the treatment they are given in the book.

The result of an economic appraisal informs the public sector investor about the economic viability of a project for society, independently of its financial returns. In addition, the linkages between financial and economic analysis include four elements that are of particular relevance to the private sector investor, as follows:

- As is mentioned above, by measuring non-monetised benefits to users, the CBA is effectively estimating the monetary value of the competitive advantage of an operation. It is an indication of the pricing power of the facility, over and above what existing prices are appropriating. The government may also look to that non-monetised benefit as a potential target for arbitrary taxes, that is, taxes meant purely to raise revenues rather than to correct price inefficiencies.
- The non-observed consumer surplus constitutes a gauge to estimate traffic that may be generated by an investment project. In this respect, the calculations involved in the economic appraisal of the investment become an input into the traffic forecast to be used in the financial appraisal.
- Differences between financial and economic costs point to possible determinants of competitive advantage that are within the power of the government to alter. For example, by conducting an economic appraisal an operator may be able to identify the cause of any abnormal traffic disparity between competing airports as being due to price distortions rather than to any inherent competitive disadvantage of the local operation.
- Finally, non-internalised external costs may signal a risk of future tax or regulatory action by the government. In the case of concessions, the investor may feel reassured by the contractual framework of the project. But contracts are a social construct, changeable if there is sufficient political benefit in so doing. Project lives spanning 20 years and more leave plenty of room for changes of government

and in government policy. The magnitude of any external costs should be a measure of the extent of the risk the operation faces.

Table 1.1   Financial and economic calculations of return on investment: differences and linkages

Item	Financial calculation	Economic calculation	Linkages
Objective	Concerned with cash flows and benefits to the private investor	Concerned with full resource use and value created to society	Sharp differences between the two may indicate: (i) desirability of financial government assistance; (ii) untapped revenue potential and the need for price regulation; or (iii) non-apparent costs
Revenues	Main source of benefits	Important source of benefits	An operating loss may hide value created to society that could only be monetised at prices that may be politically unacceptable
Operating and investment costs	Main source of costs	Main source of costs	Differences point to market distortions that may affect the competitiveness of the operation
Non-monetised user benefits	Ignored, but points to potential sources of untapped revenues	Important (sometimes main) source of benefit	A key measure of competitive advantage and potential revenue generation
Taxes	Important source of outflows	Can constitute transfers or internalisation of externalities	Can be the reason why costs differ in the financial and economic appraisals and why profit underestimates social returns
Non-monetised externalities	Ignored	Important source of costs	Non-monetised externalities signal risks of future government intervention
Subsidies	Important source of benefits	Almost always a transfer. Can also be an internalisation of a positive externality	An insufficient financial return matched by a positive economic return justifies the granting of subsidies
Interest on loans	Important source of outflows and risk (ignored here)	A transfer between owners of financial capital	Not significant
Discount rate	Weighted average cost of capital	Social discount rate or, in its absence, yield on long-term government bond	Differences between the two are due to different abilities to bear risk by private and public sectors, market distortions and government ethical considerations (mostly ignored in this book)

3 Discount Rate, Risk and Uncertainty

The rate of discount applied to estimate the present value of benefits and costs may vary between the financial and the economic analyses. The private sector financial analysis would be made with the (private) weighted average cost of capital (WACC). This is determined by the opportunity cost of equity financing, the cost of debt financing of the promoter, the promoter's capital structure and the riskiness of the project. These variables are relatively easy to observe.[2]

On the other hand, benefits and costs in the economic appraisal are discounted with the social discount rate, which at the most fundamental level depends on the social rate of time preference, the expected growth rate of the local economy and the rate of diminishing social marginal utility of income. These factors are much more difficult to measure than the components of the WACC in the financial analysis. In addition, if the size of a project is sufficiently small relative to the size of the national economy, the risk premium on the social discount rate should be removed. The estimate of social discount rate would also have to correct for taxes and other distortions in the financial markets and would need to internalise inter-generational externalities. The result is that in practice, estimating social discount rates can be a very cumbersome exercise. Unless the government publishes official social discount rates, the analyst may be better advised to rely on the real interest rate of the traded government debt security with the longest duration available as a proxy. The yield of such a security determines, after all, the marginal cost of financing of the state for long-term investment in the country.[3]

Since these issues are not specific to air transport, are widely discussed in the financial and economic appraisal literature and are largely empirical or project-specific, the issue is sidestepped

---

[2] For more on the estimation of WACC refer to textbooks on project or corporate finance. An accessible source is Brealey et al. 2008.
[3] The difficulties about what social discount rate to use can be partly sidestepped by focusing the evaluation on the internal rate of return rather than the net present value. Still, the discount rate is eventually necessary to decide on whether the estimated return makes the project acceptable.

in this book by assuming a 5 per cent discount rate on all cases whether financial or economic.[4] The subject is only briefly revisited in the discussion of economic analysis of aeronautical projects in Chapter 7.

A presentational advantage of using the same discount rate for financial and economic profitability is that the cash and non-cash magnitudes become easier to compare. This is useful, since some of the non-cash benefits and costs used in the economic but not in the financial appraisal are relevant for private investors, for example, consumer surplus. Having consumer surplus valued at a lower discount rate in the economic than in the financial appraisal may confuse private sector analysts into believing that consumer surplus is higher relative to financial profitability than it really is. For this reason, when reviewing the economic viability of investments, it may be useful for private investors to carry out the parallel exercise of discounting financial and economic returns with the same discount rate in order to gain a more realistic picture of the financial potential of the investment.

When the appraisal is based on net present value (NPV) rather than the internal rate of return (IRR), the discount rate would normally already incorporate the risk premium. Alternatively, the NPV can be estimated with the risk-free discount rate, and the reported NPV of a project would then be the risk-weighted expected value of the NPV, resulting from the probability distribution of NPV estimates.[5] This would be the normal procedure to follow in IRR-based appraisals, given that risk does not enter directly into the IRR calculation. The estimation process would usually involve three steps:

1. Performing a sensitivity analysis to see what variables have the potential to cause project profitability to diverge from the estimated central case.

---

4   Textbooks on CBA normally cover the social discount rate and its relation to market rates. Accessible sources include Boardman et al. 2014, de Rus 2010 and Campbell and Brown 2003.
5   Note that estimating a risk-weighting expected NPV from the probability distribution of NPV outcomes when the NPV has already been calculated with a risk-adjusted discount rate would amount to double-counting risk.

2. Estimating the risk-weighted expected rate of return. The resulting figure would constitute the central case, or base estimate, of project returns.
3. Estimating the probability that a project would perform below the threshold profitability below which it becomes undesirable. Deciding on both the minimum accepted level of profitability and the maximum tolerated probability of returns dipping below the threshold is a managerial decision informed by the performance of the project relative to the risk–reward profile of other investments in the sector and in the wider economy.

The mechanics of performing a risk analysis are not specific to aviation and are well documented in specialist sources.[6] Therefore this book does not illustrate risk analysis. The related issue of uncertainty, which arises where there is insufficient evidence to perform a standard risk analysis, is also covered in specialist sources on real options analysis (ROA).[7] However, in cases when there is substantial uncertainty ROA can become central to the investment appraisal. The use of ROA is illustrated in the sector chapters of this book, in two cases, including the valuation of options on aircraft, in Chapter 6, section 2, and the appraisal of innovative aeronautical projects in Chapter 7, section 2.

4   Additional Considerations

The project examples in this book show a number of simplifications in order to ease the presentation and help the reader focus on key appraisal issues. The main ones include the following:

- The estimations assume no residual value. There is no hard and fast rule about residual value estimation. Any estimate is heavily dependent on the circumstances surrounding a

---
6   For a practical guide see, for example, Vose 2008. For a summarised presentation see European Commission 2008.
7   An accessible source is Kodukula and Papudesu 2006. More technical presentations include Dixit and Pindyck 1994 and Trigeorgis 1996.

- project and the nature of the facility or technology, and ultimately rests on analyst judgement. The exception is the case of airline fleet replacement, where older aircraft are assumed to be sold.
- Long-term demand forecasting can be an elaborate process, constituting a field in itself that is outside the scope of this book. The cases illustrated in this book tend to rest on normal long-term magnitudes common in the industry.
- Prices are averaged per customer. This book does not address pricing structure, as this would entail entering the realms of industrial organisation and regulatory policy further than is already done. Any investment appraisal would have to reflect the specific regulatory circumstances of the promoter; and the investment analyst should be mindful of the implications of price regulations on investment incentives. By way of illustration, Chapter 5, on Air Traffic Management, includes an example of the types of implications that pricing policy may have for the investment decision.
- For promoters that are subject to price regulation, price adjustments tend to take place at regular intervals along the life of the project, as dictated by the terms of the applicable regime of economic regulation. The examples instead assume constant prices throughout the life of the project, consistent with any applicable regulated rate of return.
- Prices are assumed to be in real terms, that is, where inflation has been deducted.
- Taxes are simplified, applying only to inputs and outputs, rather than to profits or property. The main purpose is to illustrate the treatment of taxes on economic appraisals, rather than the effects of specific tax regimes.
- Public funds are assumed to come with no additional marginal cost resulting from the tax wedge or any other loss of efficiency.
- Finally, whereas the methods presented in the book apply to both passenger and freight transport, the presentation focuses on the passenger segment. Still, the book refers to the freight segment whenever the discussion raises issues of particular interest for freight transport.

Chapter 2
Identifying Benefits

1   Air Transport as an Intermediate Service

Economics considers air transport, and transport services in general, as intermediate services, that is, services that are used not as ends in themselves but as a means to some other ulterior consumption or production. This means that economics assumes that no one flies for the sake of flying, but to reach another location for commercial purposes, visit a friend or relative, sightsee or migrate. The implication is that transport is treated as a cost, and the passenger is understood as wanting to minimise the cost of moving from one place to another.

The cost of transport consists not only of the ticket price but also of all other elements that constitute an effort which the passenger would want to minimise. These can be summarised in the following three categories:

1. The time taken to travel from A to B.
2. The operating cost of travelling, namely the full cost of the airline ticket, which would tend to include the cost of all infrastructure, including airport and ATM charges, and the operating cost of the access and egress time taken to complete the door-to-door cycle.
3. The risk that the user takes in embarking on a trip. The cost is reflected in the user's willingness to eliminate or reduce the risk of an accident. Normally this is deemed negligible, but not in regions with poor infrastructure or services.

These three elements apply to both passenger and freight transport. In addition, in the case of passenger transport, there is a growing practice in transport appraisals to include willingness to pay to avoid discomfort, but empirical evidence in this area for aviation is less well established. The three cost categories are addressed in turn:

2 Travel Time

2.1 Measures of Travel Time

An immediate, observable component of travel time is door-to-door travel time. This would include the time to access the airport, or access time; the time spent in the airport being processed into the plane, or the departing passenger processing time; the time in the aeroplane, or flying time; the arriving passenger processing time in the airport; and the time taken to journey from the airport to the final destination, or egress time. This full sequence of door-to-door travel time, which is strongly dependent on location factors and infrastructure conditions, is already predetermined at the time the passenger buys the airline ticket.

But the passenger also experiences two additional time costs, or delays before buying a ticket. Firstly the difference between the passenger's preferred departure time and the actual time when a flight is available. This time, known as frequency delay, is reduced by increasing flight frequency. Then there is the delay that occurs when the desired departure flight is full. This delay will vary directly with load factor: the higher the load factor, the higher the likelihood that the passenger will have to travel in a different flight than the preferred flight. This delay is called stochastic delay. The summation of the frequency and stochastic delays is called the schedule delay.[1] This is a delay that is controlled by airlines when they set their schedules and their load factor targets.

Any investment in air transport capacity will affect some combination of door-to-door travel time and schedule delay. In practice, stochastic delay will largely depend on airline pricing and load factor policies, which fall within the realm of airline operations planning and are rarely affected by investments

---

1 See Douglas and Miller 1974.

Identifying Benefits 17

on infrastructure or equipment. This is especially the case in competitive airline markets, where any unaccommodated traffic resulting from high load factors by an airline may be picked up by competing airlines. The relevance of other travel time components for valuing investments will become evident in the remainder of this book. Investments aimed at enhancing passenger handling capacity will tend to affect door-to-door travel time, whereas investments aimed at increasing aircraft movement capacity will tend to affect frequency delay.

Travel time also applies to air freight. Cargo forwarders will use air transport to the extent that it pays to save time versus operating costs, such as in perishable products, or high-value manufacturers integrated in just-in-time logistic chains. The principles underlying the appraisal of time benefits for freight and passenger traffic are the same.

In practice, time-savings are often the main determinant of the benefits from aviation investments. In order to attach a value to the benefit arising from delay savings it is necessary to know how much the passengers are willing to pay for time-savings.

2.2   *The Value of Time*

Travellers are willing to spend money to save time to the extent that the time used in travelling could be used for other productive or leisure activities. How much a passenger is willing to spend to save time is called the 'money value of time', or simply the 'value of time'.

The intuition behind the valuation of time can be illustrated with a simplified example. Assume a person is offered a choice of two travel options, Mode 1 and Mode 2, to go from A to B. With Mode 1 it would take the person four hours to get from A to B and cost €50; Mode 2 would take 1 hour and cost €110. All other factors are equal. The person must decide whether it is worth paying €60 (the difference between the two ticket prices) to save three hours (the difference between the time taken between the two modes). If the answer is yes, it implies that the person's value of time is at least €20 per hour (€60 divided by three hours), meaning that the traveller is willing to pay at least €20 in order to save an hour. If the answer is 'no', then the traveller's value of time is less than €20.

Studies of value of time gather evidence on many such choices to compute statistically significant monetary values, normally expressed in currency units per hour. The value of time is determined by many variables. A key one is income, with a direct relationship between the two variables. The cases presented in this book assume that the value of time grows on average at 2 per cent per year, in line with the assumed growth in income per capita.[2]

Other factors include trip purpose. Generally, working time, leisure time and commuting time are valued differently, with evidence showing higher willingness to pay for working travel time. Value of time research is widespread and many estimates are available in official government guides and the academic literature. A widely known example is HEATCO (2006), a research project sponsored by the European Commission, which finds an average value of time in the European Union (EU) for airline business travellers of €32.80 per hour and €13.62 for long-distance leisure travellers, both at 2002 prices. The latest available guidance from the US Department of Transportation at the time of writing was for US$57.20 per hour for business trips and US$31.90 per hour for personal trips.[3] It should be noted that both papers include also guidance regarding margin of error.

The willingness to pay for time-saving in the cargo sector derives from elements such as perishability of the product,

---

[2] The relationship between income per capita and value of time comes hand in hand with the relationship between income per capita and labour costs. In principle, labour costs should also grow with income per capita, increasing the unit costs of a project. On the other hand, growth in income per capita generally implies growth in labour productivity, decreasing unit costs. The important thing for the investment analyst is to bear in mind that when making assumptions about growth in the value of time over the lifetime of a project, the analyst should also make assumptions about growth in labour costs and productivity. If the analyst assumes that value of time grows in real terms over time, but assumes that labour costs do not, the analyst is implicitly assuming that there are sufficient productivity gains to compensate for the growth in labour costs.

[3] See Belenky 2011. The DoT recommends equal values of time for high-speed railway and air travel.

value-to-weight or value-to-volume ratios of the product, and the time sensitiveness of just-in-time production chains. Estimates on the value of time for air freight are difficult to come by, let alone estimates disaggregated by product perishability. Unfortunately HEATCO (2006) quotes freight values of time for road and rail transport only. There is, however, indirect evidence from the trade literature.[4]

3 The Money Cost of Travel

3.1 The 'Out of Pocket' Money Cost of Travel

The money cost of travel involves the 'out of pocket' money price that the traveller pays for the door-to-door journey. This includes the operating costs and any return on capital for all operators involved in the door-to-door transport chain, including the airline, and whichever means of transport the passenger uses on the airport access and egress journeys. Normally, payments to infrastructure providers are included in the ticket price, but if they are not, they should also be included in this category.

Together with the time invested in the journey, the money cost to the traveller constitutes the key parameters in shaping the decision of the traveller on what trip to make and whether the trip is made at all.

When making an economic appraisal, however, there are additional considerations to take into account, due to the frequent presence of distortions in the money cost of travel.

3.2 Distortions to the Money Cost of Travel

The money prices the traveller pays may not reflect the opportunity costs of the resources employed in producing the transport services purchased. Taxes, subsidies, externalities and imperfect competition can result in prices diverting from

---

[4] Hummels (2001), for example, approaching time as a trade barrier, finds a time cost per day equivalent to a 0.8 per cent ad valorem tariff for manufactured goods. Other similar studies include Nordas 2006 and Hummels and Nathan Associates 2007.

resource opportunity costs, in which case the economic appraisal will need to make adjustments to observed money prices.

Figure 2.1 illustrates the case of a tax on an input. The supply curve of the input is depicted by S+t, including a unit tax equal to t over the tax-free cost of the input. S would correspond to the distortion-free supply of the good, reflecting the input's opportunity or economic cost p_e. The tax causes an undersupply of the good equal to $q_{e1}-q_1$. Let us say that the project causes demand for the input to increase from D_1 to D_2, shifting quantity demanded from q_1 to q_2. The observed price remains constant at p+t, which would be the cost that the project promoter uses in the calculation of the financial returns of the project. However, this financial cost disguises a welfare gain, resulting from an increase in the supply of an under-supplied good, equal to area abde, which is transferred to the government through the tax on the input. An economic analysis would have to deduct that welfare gain from the observed input costs q_1abq_2, resulting in an economic cost of the input equal to q_1edq_2. This alternative price reflecting economic costs is called the 'shadow price' of the input, to distinguish it from the out of pocket or observed price.

As is implied by Figure 2.1, economic cost considerations are of no relevance to the passenger, who will make the travel decision following observed out of pocket money prices, independently of how distorted those prices are. Therefore, when conducting the economic or the financial appraisal of an investment, consumer behaviour is inferred from out of pocket prices and not from shadow prices. So the economic analysis would use the quantities observed in the financial analysis, which are, after all, the actual quantities of goods supplied to the project, but would value them at p_e, rather than p+t.

The most common adjustments involve energy costs, should energy taxes apply, and labour costs. Taxes and net social security contributions can be viewed as transfers and would then be removed from costs, to produce the shadow price of labour, or the 'shadow wage'. The last section of this chapter includes a fuller discussion of employment issues.

Identifying Benefits

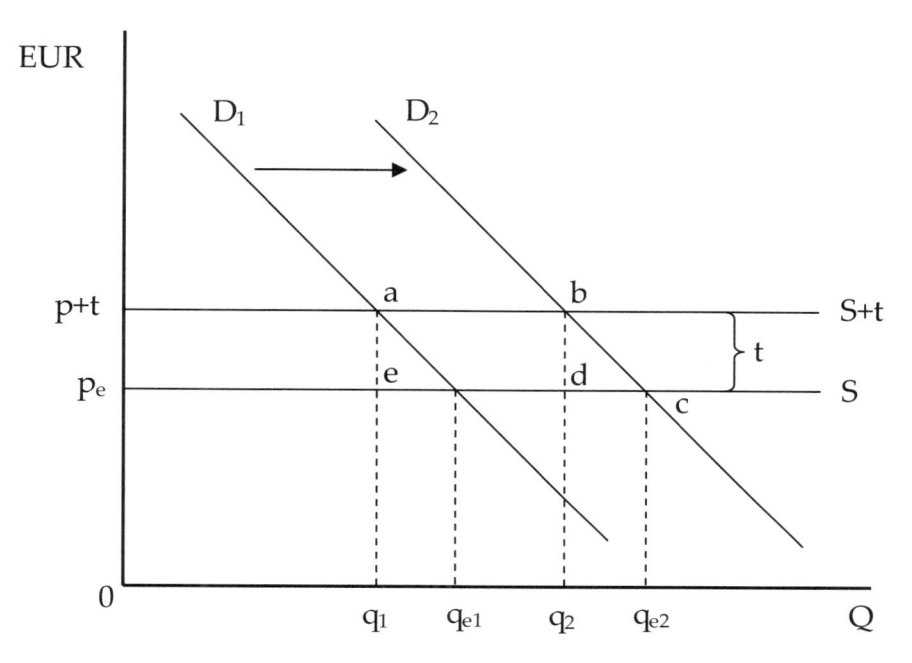

Figure 2.1 Effect of a tax on the money cost and economic cost of an input

Subsidies would have the opposite treatment. In the financial analysis a subsidy constitutes an income for the project or a saving to investment or operating costs. In the economic analysis that apparent income or cost saving is recognised as a transfer of resources from the government to the project, and as such must be added back to costs.

Developing countries may witness a wider set of distortions, including, for example, capital controls and wide price differences between formal and informal sectors. Shadow prices are well covered in the economics literature, and the reader is referred to those sources for a fuller discussion.[5]

Finally, it is worth pointing out that whereas such adjustments are not necessary for the estimate of financial return, they still offer useful information relating to the financial return of a

---

5   See, for example, de Rus 2010 and Campbell and Brown 2003 for an introduction; and Londero 2003 for a fuller treatment.

private investor. They point to areas where government policy may be inefficient and arbitrarily affect travel choice. They may point to areas of potential future changes in government policy with adverse or favourable implications for money prices.

4 Accident Risk

People are willing to pay to reduce the risk of serious injury or loss resulting from accidents. Likewise, freight forwarders buy insurance against loss of or damage to a shipment. In addition, accidents incur medical and legal costs, as well as loss through damage to equipment and property. Transport projects therefore generate benefits if they bring about reductions in accident risk, safety improvements being a legitimate component of any transport project appraisal. However, for normal aviation investment projects in countries with well-developed institutions, such benefits are very small relative to other benefits.

Project benefits resulting from reduced accident risks can accrue in two respects. The first is through the improvement of safety conditions within the existing transport mode. In countries with well-developed institutions, aviation operations are not allowed to take place if they do not comply with safety standards, and such rules leave the risk of death or serious injury very low. In this respect, the case for an aviation investment aimed at bringing a facility to meet safety standards does not depend on the benefits of increased safety. If the investment is not made, the facility cannot operate at all. Therefore, the investment decision will depend on whether the necessary investment cost to meet safety standards leaves the facility still viable or not.

Ironically, therefore, in practice the value of safety is of no (direct) relevance in the appraisal of investments aimed at improving safety. Instead, the economic analysis of safety measures becomes relevant in deciding what the safety standards should be. In that case, the analysis would enter the field of economic appraisal of policy, not of economic appraisal of investment projects. Once a policy is set, safety standards will determine project design characteristics and costs, and through such costs affect individual project viability. In this respect, safety-related costs and benefits are always (indirectly) present in project appraisals.

The second respect in which safety benefits can accrue to projects is in situations when a project causes the shifting of traffic between modes, which, while meeting regulatory safety standards, have different safety records. Road transport has generally a higher accident rate per passenger-kilometre than air or rail. Road transport does not, however, provide a suitably close substitute for air transport. The closest substitute, but for short-haul trips only, would be high-speed rail. But the safety record of high speed rail is not dissimilar to that of air transport.

Road transport becomes relevant for air transport appraisals mainly when passengers switch between airports by road. But there again, assuming normal driving conditions, the benefit that aviation projects generate by avoiding road accidents is a small proportion of the broader benefits of the project. This is illustrated in the discussion of a greenfield airport, in Chapter 4, section 1.

Therefore, safety generally plays a minor role in justifying air transport investment.[6] In projects with a weak institutional framework, where international safety standards are not met for air and other transport modes, investments in safety gain a higher prominence among project benefits. Such situations, however, are not covered in this book.

5 Externalities

As seen in the preceding section, user prices can be distorted through mechanisms such as taxes and subsidies. However, they can also be distorted by failing to reflect costs imposed on parties not directly involved as consumers or producers in an air transport project. Ultimately this is due to poorly defined property rights. In any free market, the production and sale of any good or service is the result of the free trading decisions

---

6   See Chapter 4, section 1.

of the direct participants in the transaction, namely the consumer's decision to buy, and the producer's decision to sell. Such decisions are based on the costs and benefits perceived by each of the two parties, and will depend on such participants finding the transaction mutually beneficial.

Welfare economics argues that transactions are worthwhile to society when they result in a net gain to social welfare, which occurs when the value of a good or service to the consumer is higher than the value of the resources used up in its production. When both parties involved in the transaction freely agree to transact, the transaction can be expected to be beneficial to society, that is, to result in an improvement in social welfare.

But the transaction may result in costs to third parties that are not involved in the decision to use or supply the service. Such parties may not have a legal entitlement to claim compensation for the cost or damage incurred. Examples may include people who have to endure aircraft noise, or who must be relocated to allow the building of airport facilities. Indeed, the 'third party' may consist of large parts of the world population, which may experience costs from global warming resulting from greenhouse gas emissions.

Advanced societies increasingly grant de facto property rights by means of compensation, taxes and other restrictions on production or consumption in order to ensure that the primary participants in the transaction – the buyer and the seller – include costs to third parties in their decision to transact. Examples include fuel taxes (not applicable to aviation at the time of writing), requirement to buy emissions rights, and noise-related landing charges at airports. When that happens, third party, or 'external', costs are said to be 'internalised'. The resources raised can be used to finance compensation such as installing double glazing in properties affected by noise, financial payments for relocation, or investing in carbon-capturing sectors such as forestry. The result is that whereas the transaction takes place only if it is mutually beneficial to the buyer and the seller, the economic calculation involved in the trading decision includes costs to third parties, so that the transaction can be taken to constitute a welfare improvement for society.

In cases where externalities are internalised, the financial return of an investment already includes the external costs. Moreover, the amount of goods consumed and produced will reflect such costs. Assuming there are no other distortions, the financial return of the project also reflects the economic return.

However, when externalities are not internalised, the financial analysis does not reflect externalities. Hence, the economic appraisal of the investment should include the external costs as additional to the financial calculation. This will rely on the availability of data on relevant shadow prices for the externalities concerned. The main environmental externalities of aviation include greenhouse gases, contrails, noise and air particles. The academic literature offers many estimates, but academic papers tend to offer location- and method-specific results. It is therefore more prudent to use studies that amalgamate results from a number of papers. Examples of this type of study include CE Delft 2002 and HEATCO 2006.[7]

It should be noted that since quantities produced and consumed will not be affected by such external effects, consumption and production will be higher than if such costs had been internalised. The resulting economic costs would tend to be of a greater magnitude than when they are internalised.

The economic analysis is playing a dual role. It helps the public sector planner measure the actual returns to society of the project. And it helps the private sector analyst by pointing to areas of risk for the promoter regarding future government intervention. However, whereas the economic analysis identifies the risk and measures the potential cost, the actual cost to the promoter of an eventual government intervention to internalise the external cost would depend on the precise policy instrument the government decides to apply.

This raises a possible scenario of government intervention aimed at other objectives, but resulting in similar outcomes as intervention aimed directly at internalising an externality.

---

7   For a broad discussion of air transport and the environment see Daley 2010.

This may occur with items such as arbitrary air passenger charges levied uniquely for money-raising purposes. If the result of the arbitrary charge is raising the price of airline tickets by an amount at least as large as that which would result from internalising an existing externality, as far as the economic analysis is concerned the air passenger charge may fully offset the effects of the externality.

It is worth highlighting that externalities do not concern only costs, but can also constitute benefits, such as when an aviation project helps alleviate road transport congestion, or creates knowledge that can be used in other industries.

Also, beyond externalities, aviation investments can bring about benefits to third parties through price effects on secondary markets. Aviation services can generate substantial economic activity in the region where they are located and enable the exploitation of economies of scale for certain products. As discussed below in section 7.3, these are valid indirect economic benefits to be attributed to a project.

When an aviation project benefits third parties through externalities or indirect benefits, aviation investors may enlist the third parties likely to benefit from the project to support the investment. This is another significant piece of project information generated by an economic appraisal that may be important to management, and which is not captured by the financial appraisal.

6 The Generalised Cost of Transport

The total generalised cost of transport adds up all the costs involved in transportation for the user and for society at large. A distinction is normally made for the subset of costs that are borne by the user and which therefore determines travel behaviour, called the 'behavioural generalised cost'. The total generalised cost would also allow for any subsidies, externalities and other distortions. In this book the term 'generalised cost' is used to refer to behavioural generalised cost for reasons explained later in this section. Subsidies, externalities and other distortions are included in the analysis separately. An example of (behavioural) generalised cost calculation is included in Chapter 4, section 1, and a simplified example distinguishing between behavioural and total generalised cost in Chapter 6, section 3.

As is explained at the beginning of this chapter, transport is an intermediate good, and the transport consumer will try to minimise the cost incurred in travelling between points A and B. As far as the user is concerned, the value of an investment in a transport facility will be measured by the extent to which it reduces the generalised cost of the user when making the trip. In making a travel decision, transport users will consider all options available: transport modes such as boat, rail, car, or air, and within air, all routings, operators and alternative departure and arrival airports available. Indeed, if the option yielding the least generalised cost is deemed too high, the prospective passenger will decide not to travel.

The measure of time included in the generalised cost of travel would normally be the door-to-door travel time plus the frequency delay. Whereas frequency delay is harder to measure than door-to-door travel because it depends on subjective departure time preferences, it is still an important driver of traveller behaviour and willingness to pay. Lack of sufficient departure frequency can be a reason to travel through alternative airlines, airports and modes of travel, or not to travel at all. However, frequency delay becomes relevant to investment appraisal in situations where the project affects departure frequency, otherwise similar delays with and without the project means that the frequency delay cancels out in net terms. For the reasons mentioned in section 2 above, stochastic delay is left out of the analysis. An example of dealing with frequency delay is included in Chapter 4, section 7.

Another component of generalised cost would be discomfort and the willingness to pay to avoid it or minimise it. The higher ticket price of business class seats is not an adequate measure because it mixes comfort issues with ticket flexibility. Also, frequent flier programmes introduce principal-agent issues. Evidence for willingness to pay for comfort factors and service attributes is emerging.[8] However, the evidence so far is mixed, and additional research would be required before estimates can be incorporated reliably as a welfare consideration. Over the last decade there has been a growing application of stated preference techniques to model air travel demand, enabling the study of variables that had been harder to model with revealed preference techniques.[9] Hopefully, a growing literature of published studies will cast more light on passenger willingness to pay for service attributes over the next few years.

As a factor to weigh on the decision to invest in air transport at all, comfort really is relevant on short-haul trips where the traveller faces competing transport modes. On long-haul trips, where the only choice is air transport, comfort conditions with and without the project are on average the same, and comfort becomes an issue of inter-airline competition, rather than one of whether to invest in air transport at all. Still, even on short-haul trips the level of comfort offered by airlines and high-speed rail is comparable, and choice of travel mode tends to be made largely on travel time and money cost. That is, any net benefit contributed by comfort issues is likely to be dwarfed by other components of generalised cost.

Turning to the total generalised cost, in addition to allowing for subsidies and externalities, it would use economic or shadow prices to measure the scarcity of resources (see section 3.2 above), instead of out-of-pocket prices. In this sense, the total generalised cost can be thought of as the *economic* generalised cost, as it would measure the actual resources used up by the traveller, to distinguish it from the subset of costs that would constitute the *user* or behavioural generalised cost described above, which corresponds to costs incurred by the traveller. The mechanics for calculating investment return would vary slightly depending on the measure of generalised cost used. Using the total or economic generalised cost would still require an estimate of the user generalised cost in order

---

8 See, for example, Hess et al. 2007, Ling et al. 2005, Lu and Tsai 2004, and Coldren et al. 2003.
9 See Garrow 2010.

Identifying Benefits 29

to make demand projections. As mentioned above, this book focuses on behavioural generalised cost. It links generalised cost to observed demand, making the appraisal exercise more intuitive and the financial appraisal easier, and enables the use of the same generalised cost measure in both the economic and financial analyses. The economic analysis then makes the necessary adjustments to the financial analysis to include all other effects.

Table 2.1 sums up the components of generalised cost as will be used through this book. It is important to highlight that, as will become evident, for investment appraisal the relevant magnitude is the change in generalised cost brought about by the project relative to those costs that users would face without the project, rather than the absolute generalised cost.

Table 2.1   Components of generalised cost of transport as used in this book

Cost item	Usage
Travel time: door to door	Included
Travel time: frequency delay	Included
Money cost of travel	Included
Safety	Included, but significant mostly in situations where travel conditions are particularly unsafe.
Comfort	Excluded, effect deemed to be relatively small.
Externalities	Excluded from generalised cost but added to the economic analysis as additional costs.
'Shadow price' adjustments to observed prices	Excluded from generalised cost, but included in the economic appraisal as a separate adjustment.

7   Wider Economic Benefits

Analysis of economic returns from transport investments often include among project benefits items such as multiplier effects, tourist expenditure in the local economy, job creation and increases in the value of land. They constitute secondary markets (where the primary market is the transport market that the project addresses) and include all markets that will feel the impact of the project. All these effects are intuitively appealing and often reflect actual benefits of the investment. However, there are two problems affecting their inclusion in the economic analysis of the investment. First, many of them double-count benefits already picked up by savings in the generalised cost of travel. And second, whereas some may measure actual benefits, they do not measure incremental benefits and do not take into account the alternative use of resources in the absence of the project and, therefore, do not constitute appropriate measures to guide investment decisions.

Ultimately, the standard economic appraisal techniques – focusing on changes to full or economic generalised cost of transport, measuring inputs at opportunity costs, and including externalities, as set out in the chapters that follow – measure the full benefit of the project to the local, national and world economies. Consideration of secondary markets is the exception rather than the norm, as is explained below.

The rationale behind this conclusion rests on the economic information that prices reveal under different market circumstances regarding competition and distortions. The discussions that follow apply to all economic appraisals in transport and other sectors. They are not particular or specific to air transport. Therefore this section includes only a brief summary of the key arguments. The reader is referred to the specialist literature for a more detailed treatment.[10]

7.1 Prices Reflect Marginal Valuation and Opportunity Costs

The valuation of a user for a good or service is revealed by the user's willingness to pay for them. Looking at the economy as a whole, at a given point in time consumers will spend their income on the combination of goods and services they prefer

---

10 See, for example, de Rus 2010. For a broader treatment of transport investment in economic development see Banister and Berechman 2001. The reader should bear in mind the distinction to be made between the appraisal of economic viability, which measures changes in welfare, and impact analysis on income or employment, regardless of the net effect on welfare. See below, sections 7.5 and 7.6.

most (that is, that maximises their utility). Inter-temporally, they will borrow or save according to their preferences for present over future consumption and the prevailing interest rate. Meanwhile, producers will compete to produce with the most efficient available technology to satisfy customer requirements, and through competition will end up supplying their products at normal profits (which will be equal to the risk-adjusted interest rate). That is, consumption and production in the whole economy are solved simultaneously to yield the combination of goods and services most valued (welfare maximising) by consumers, for a given state of technology and resource availability.

When this happens, any observed pattern of consumption and production reflects marginal consumer preferences (including valuations of the range of products available) and marginal production costs (that is, price equals marginal cost). For any additional good or service to be produced, it must be marginally more desirable than the alternative use of resources, and it will be produced with a normal profit. Hence that marginal unit produced must be valued at the margin, namely as its observed money price.

According to such reasoning, in a competitive market, without distortions, the observed financial profitability of a given investment project reflects normal risk-adjusted profits resulting from efficient production and the price at which the output is sold reflects marginal valuation. Therefore, in such market circumstances the financial profitability of the project is taken as a fair reflection of economic profitability.[11]

This is the underlying assumption that is applied to those sectors that are deemed highly competitive, such as airlines and, perhaps to a lesser extent, aeronautics. In reality, markets in those sectors still present some distortions, mostly taxes, subsidies and distortions on secondary markets. The investment appraisal will need to make adjustments, as will be shown

---

11 See Varian 1992 for a formal proof. Note should be taken that in a project appraisal context income to production factors have an opportunity cost. Therefore in a perfectly competitive economy marginal projects would tend to have an economic net present value of zero.

in the cases examined in later chapters, but the financial and economic returns will tend to be relatively close.

7.2 Differences in Generalised Cost Reveal Value

Unlike the airline and the aeronautical sectors, the supply of infrastructure services tends to be far from highly competitive. Indivisibilities in capacity provision mean that marginal increases in capacity may be lumpy, giving rise to both sunk and fixed costs. This means that infrastructure operations will exhibit strong cost economies and minimum efficient scales that render the sector prone to monopolistic outcomes. Therefore, when an airport suffers congestion, the alternative airport may be some non-trivial distance away. In those circumstances, the user will experience costs in switching to the alternative airport, possibly involving additional hours of travel. The switching cost to the user is measured by the difference in the generalised cost of transport between the alternatives, and the user will be willing to incur it to the extent that the traveller still values the trip highly enough. The switching cost therefore reveals consumer surplus available from using the preferred airport.

In other words, the cost of switching from facilities A to B measures the additional value that A is creating to the user relative to B. To illustrate, say airport A is congested and does not have airline seats left to the desired destination. The user has to drive to an alternative airport B located two hours drive away, incurring an additional generalised cost of, say €60, relative to the generalised cost experienced when travelling through A. Then, those €60 measure the user's additional willingness to pay for additional capacity at airport A, and therefore measures the (incremental) value that airport A offers.

Users can consist of passengers or shippers. For freight shippers, the transport will almost always be a component of a production chain. In the case of passengers, trip purpose can either be leisure or business. For leisure users, the generalised cost is an element of the total valuation of the final good (say, a holiday). The leisure traveller's willingness to pay to

Identifying Benefits 33

avoid switching costs will ultimately depend on how much the traveller values the holiday.

For business or work-related travel, as for freight, the ticket price is an input cost in the production chain. Businesses will be willing to incur the cost to the extent that it ultimately produces a good which is valued by the final user sufficiently to make the trip worth it. This same consideration also applies to the willingness to devote paid worker time to travelling. The value of the time invested in travelling must ultimately reflect the value of the output produced as a result of that trip. That is, a business will invest the time of its workers in travelling to the extent that it is profitable to do so. And the worker's revealed value of time (in other words, the amount the firm will be willing to pay to save worker travelling time) will reflect the value of the output that that worker could have produced with that time, that is, the worker's time opportunity cost. That is to say, working value of time will measure the opportunity cost of output foregone and, by implication, the money valuation of the time savings yielded by a transport project will reflect the amount of additional output enabled by the project.

The implication of the above is that the savings in generalised cost that a project grants to local businesses reflect the value that the airport generates to the local economy in terms of enabled additional production. This implies that, in economic appraisals, savings in user generalised costs already reflect production benefits in the local economy, so that adding additional benefits to firms would constitute double counting. There is an exception to this conclusion, though, discussed in the next section.

7.3 Secondary Markets

Economic appraisals of investment projects incorporate additional output-related benefits to secondary markets depending on whether a project affects observed prices in the secondary market or not. The difference is illustrated with the help of Figure 2.2. Let us assume that the secondary market in question consists of engine lubricants and that the market for lubricants is free of distortions. An airport project will cause the demand for lubricant products to increase in the local economy. The market for lubricants is large – so that economies of scale have been exhausted – and the market is competitive, with the marginal cost (and supply) curve as depicted by S_a. The increase in demand for lubricants brought about by the project is illustrated by the shift from D_1 to D_2, increasing the quantity of lubricants demanded from q_{a1} to q_{a2}, but this has no impact on prices, which remain the same before and after the project ($p_{a1}=p_{a2}$). There is no impact on marginal costs either and suppliers of lubricants continue making normal profits. The implication is that the scarcity of lubricants in the local economy is left unchanged by the project. The project has no knock-on effect on the local economy that affects the welfare of third parties, other than through distortions such as taxes or externalities associated with the lubricants market.

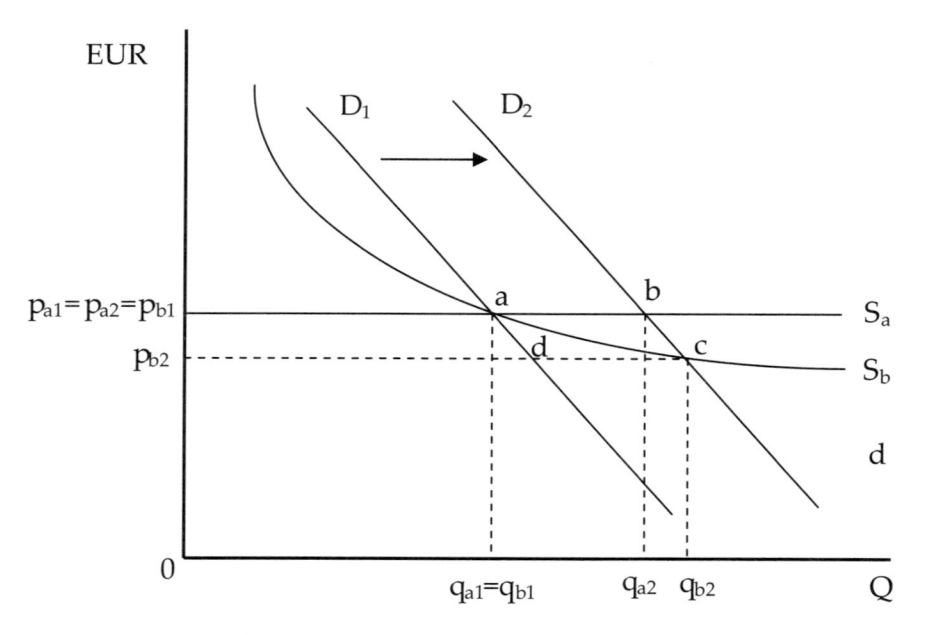

Figure 2.2   Effects of a project on secondary markets

Identifying Benefits

Assume instead that the market for lubricants is small, still enjoying economies of scale, and that the project brings about a substantial increase in market size, allowing suppliers to exploit economies of scale. This would be the situation illustrated by the marginal cost curve S_b.[12] The project shifts demand from D_1 to D_2, causing quantity demanded to increase from q_{b1} to q_{b2}, but now the price of lubricants in the local market falls from p_{b1} to p_{b2}.[13] The airport, the airlines operating from the airport and other suppliers of airport services will enjoy a lower price of lubricants than was the case before the project. This constitutes a primary market benefit which will show in the standard calculations of financial returns (the airport) and the economic returns (the airport, airlines and eventually passengers) of the project.

However, other users of lubricants in the local economy (for example, factories and road hauliers unrelated to airport activities) will also enjoy lower prices. The project has made lubricants less scarce in the local economy. This brings about a welfare gain in the secondary market equal to the area p_{b1}-a-d-p_{b2}, which the financial analysis will ignore, but which the economic analysis will have to include as a knock-on benefit of the project to a secondary market (the local lubricants market).[14] This production

---

12 This illustration abstracts from the implications of the declining marginal cost curve for the structure of the secondary (i.e. lubricants) market. With increasing returns to scale the market will not be perfectly competitive. Instead there would be some alternative, less efficient structure such as monopoly or possibly some form of cooperative oligopoly. This would have implications for the estimation of welfare changes resulting from the project, depending on the extent to which the cost savings are passed on to users or appropriated by the producers of lubricants. This illustration only introduces generically the types of situations where a project may have welfare implications for secondary markets.

13 This assumes that cost savings in lubricant delivery are passed on to users of lubricants. The conduct of the suppliers of lubricants may be different, with implications for the extent of welfare changes, as suggested in footnote 12. Note, however, that even if lubricant suppliers appropriate all of the cost savings, their increase in profits constitutes a welfare gain. Other things being equal, such a gain in welfare would be less than the corresponding welfare gain had the market for lubricants enjoyed marginal cost pricing.

14 Area p_{b1}-a-d-p_{b2} also includes lubricants usage by the airport without the project. Care must be made not to double-count this benefit.

benefit is not picked up by valuations of saved time and would have to be added as an extra benefit to the local economy.

The effect of the project on a secondary market could also be adverse. In cases of decreasing returns to scale, implying an upward supply curve, the project will increase prices in the secondary market, bringing about a knock-on welfare loss to the local economy that must be subtracted from the economic returns of the project.[15]

To conclude, effects on secondary markets should be taken into account when they bring about a change in prices in the secondary market, and should be ignored otherwise.

However, yet again, there is an (apparent) exception to this rule. In the case of transport projects, changes in land prices in the local economy brought about by the project require some additional considerations.

7.4 The Value of Land

There is a close relationship between the value of property and its proximity, or accessibility, to desirable locations, such as a city centre, a high-quality residential area, a beach, or a centre with economic activity. Improvements in transport services in an area enhance the accessibility of the area. Airport projects, like any other transport infrastructure development, tend to increase the value of land in their vicinity. The exception would be those areas affected by negative externalities of a project which, in the case of airports, consist mostly of those areas below noisy landing and take-off paths.

The extent to which a transport facility is desirable will be reflected in the amount of traffic the facility processes. People and firms will relocate to an area close to an airport to the extent that they or their clients use the airport, and their willingness to pay for property in the new location will be commensurate with how much they value the improved accessibility supplied by the airport. Improved accessibility can be measured through savings in generalised cost of transport enabled by the airport.

---

15 Care must be made not to mix a welfare loss with an increase in producer surplus that may accompany such a scenario.

So, users who value proximity to the airport will relocate to the airport vicinity and will be willing to push property prices up to the present value of its expected savings in generalised transport costs. Meanwhile, local residents in the vicinity of the airport who do not value proximity to the airport by as much will sell their properties to those who value such proximity to the airport. In effect, those selling their property are appropriating the buyers' capitalised value of the improved accessibility to the airport. The increase in the value of the property therefore constitutes a transfer, rather than a generation of value additional to the savings in generalised cost of transport produced by the project. The implication is that including land price increases resulting from a transport investment as a benefit of the investment will double-count benefits that are already being included in the analysis through savings in generalised transport cost.

The previous section of this chapter, 7.3, showed how the effects of a project on secondary markets should be ignored when the project does not bring about a price change in the secondary market, but can be the source of additional benefits when the project does change prices (or suffers from price distortions or imperfect competition). Land is a secondary market for transport projects and the conclusion stemming from this section – that increases in land price should not be counted – may seem to contradict the conclusions reached in section 7.3 of this chapter. In fact there is no contradiction because here, no claim is being made that the changes in land prices are not a benefit; rather, it is suggested that they are already taken into account by changes in generalised transport costs produced by the project.

The preceding section, 7.3, showed that the economic analysis should include benefits arising from price changes in secondary markets, as these are not reflected in the generalised transport cost savings of the project. In the case of land, the changes in land prices are the capitalisation of the changes in generalised costs, and they are already reflected in those changes in generalised cost. Indeed, in a hypothetical project where there is poor data on both the value of time and the origin of trips to the airport (that is, where it is not possible to compute savings in generalised transport costs resulting from the project), changes in the value of land can be taken as a surrogate measure of the accessibility benefits brought about by the project.

Just as increases in land prices measure capitalised benefits, falls in land prices measure capitalised losses. It is mentioned above that aircraft noise can bring about a decline in property prices in affected areas. Such a decline would be a surrogate measure of the noise externality, and not an additional cost to a monetised measure of noise externality in an economic appraisal.

The investment analyst should proceed with care in gauging the expected increase in land prices resulting from a project. If the analyst is appraising a project after it has been announced to the general public, it may well be that land prices already reflect at least part of the expected benefits of the project.[16]

7.5 Multiplier Effects

Income multiplier effects resulting from expenditures in project inputs and from project outputs do not take part of appraisals of economic viability. This is because had the funds invested in an alternative use they would also have caused multiplier effects. Any expenditure will generate multipliers. Even projects that both lose money and generate a net welfare loss will still generate multipliers.

The net difference that the project will make to income and welfare consists of net monetised and non-monetised value generated, which is what cost-benefit analysis measures. Multipliers are the domain of impact studies, which describe the effects of a project on the local economy, but do not address the question of whether the project generates a profit or a net welfare improvement over and above the opportunity cost of inputs.[17]

---

16 The possibility that land prices start to increase before a project is announced to the public should not be ruled out, particularly in conditions of poor institutional quality.
17 See Crompton 2006 for a discussion of misuses of multiplier effects within a travel context.

7.6 Job Creation

Another common source of error in economic appraisals is the treatment of employment. Whereas job creation is good and is welcomed, it is very common to cite job creation as a justification for an investment project. On the other hand, there is no need to explain to business people that labour constitutes a cost. Labour is a scarce productive resource. Occupying a worker in a project precludes other businesses from employing that worker. Therefore, subject to the frequent distortions in labour markets (such as taxes, social security contributions and restrictive labour market laws), salaries reflect the opportunity cost of labour, a scarce service.

The opportunity cost of labour can be illustrated with a simple example that reminds us that countries become richer when a task can be done with less labour input (increasing labour productivity), freeing labour resources for other tasks. If society can make a B-747 fly with three pilots (two in the cockpit plus one in reserve) instead of four (three in the cockpit plus one in reserve) society will be richer because it can create a service (flying a B-747) with fewer resources (labour input), releasing a pilot to operate other flights.

However, whereas labour is an input, as we saw above in section 3.2, input costs can be distorted due to the presence of taxes or subsidies. In the calculation of economic returns labour taxes and social security contributions should be deducted from the money labour costs to estimate the shadow price of labour. In that sense, there is a 'benefit' to using labour inputs in a project in the form of a deduction from project costs of what is in fact a transfer to the government or to a social security fund.[18] Such a 'benefit' works out to a lower input cost, rather than a net benefit.

---

18   A less orthodox but perhaps more pragmatic approach would be to view labour taxes and social security contributions as necessary payments for the good functioning of society, like paying fire insurance for buildings. That would save the analyst from having to estimate shadow wages. In any case, such adjustments are rarely of sufficient magnitude relative to other project costs to make a significant difference to the outcome of an investment appraisal.

In addition to taxes and social security contributions, shadow wages can also correct for additional distortions to the labour market, such as high unemployment benefits, the existence of minimum salaries and highly rigid labour market laws that may result in unemployment.[19]

---

19 See Londero 2003.

Chapter 3
The Basic Framework

Introduction

Aviation investment projects may be classified into two broad categories: landside and airside. Landside investments would involve projects that enhance the capacity of the system to process passengers or freight, in terms both of quantity and quality. Airside projects are those that expand the capacity to handle aircraft, in terms of number of aircraft movements or aircraft size or take-off weight.

This chapter introduces the underlying conceptual models used to evaluate the returns of investments in airside and landside projects.[1] It also discusses issues that arise when building the 'with project' and 'without project' scenarios for appraisals.

1   Landside Investments

Landside investments concern the quantity of passenger and freight throughput in the air transport system, and the quality (or level) of service offered to those transport users. This section of the chapter discusses passenger transport, but the framework applies equally to freight transport.

The market for air travel can be modelled as described in Figure 3.1. The graph presents the case of an airport, but could also be used to describe airline and air traffic management (ATM) investments.

---

1   A formal presentation of the models that follows is available in Jorge and de Rus 2004.

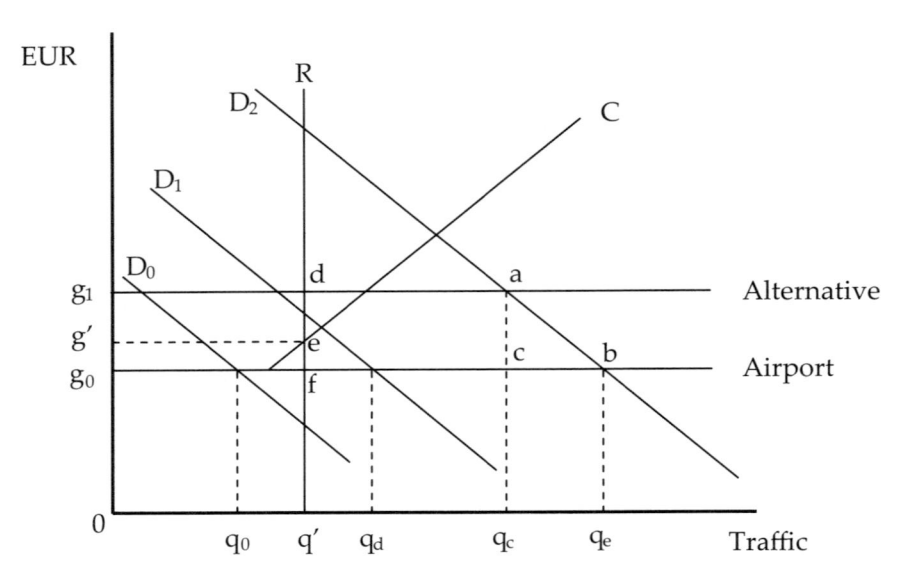

Figure 3.1 Demand and supply in landside capacity provision

For a certain region, g_0 is the generalised cost to the average traveller of using the local airport; this is referred to from now on as 'the airport'. The generalised cost to the same travellers of using the immediate alternative means of transport would be g_1. This alternative could be another airport located outside the region. Note that the fact that the generalised cost to the customer differs between the alternatives implies that the market is characterised by product differentiation. At the extreme, when the product differentiation is very large, the situation would be akin to a monopoly. The current situation therefore does not reflect perfectly (or highly) competitive market conditions. The implication of this issue for scenario-building is discussed in section 3.2 below.

The analysis proceeds by considering the demand for which the airport represents the preferred means, or node, of travel. When demand conditions faced by the airport are as described in schedule D_0, traffic at the airport would be q_0. The airport can accommodate all passengers with an average generalised cost of g_0. As demand grows the demand curve shifts rightwards. At some point it will reach design capacity. From that point onwards, further demand growth will cause congestion in the terminal, creating time costs to travellers in the form of higher passenger processing times, greater likelihood of departure delays, or forcing them to travel at less preferred times. As demand grows, the cost to the average traveller will also grow, as denoted by schedule C. The airport will generally establish a certain capacity level beyond which it will begin rationing capacity and negate airlines additional check-in desks, larger boarding gates or slots. This rationing is depicted by schedule R, taking place at a throughput of q' passengers, who would experience a generalised cost of g'. The difference in money terms between g' and g_0 along the vertical axis measures the cost of congestion to the average passenger using the airport when the airport is at capacity rationing stage.[2]

Should the airport choose not to ration capacity, as demand grows, shifting the demand schedule rightwards, airport throughput would increase beyond q', making the airport increasingly congested. At some point, congestion, and the accompanying passenger generalised cost, would reach a point where the average passenger would be indifferent between using the airport and the alternative means of travel. This situation would be depicted by the intersection of the curves

---

2   This congestion cost is incurred by the users of the facility. However, in network industries, like air transport, underperformance in one node may cause disruption in other nodes, generating further costs to users elsewhere in the network. Any such costs caused by the project under appraisal should be incorporated into the calculation of economic returns. Notionally, they can be treated as the project affecting costs on secondary markets, discussed on section 7.3 of Chapter 2. The magnitude of the delay is project specific and relies on models of the network surrounding the airport. The US FAA has long recognised the relevance of system-wide delay propagation (see FAA 1999) and has been pursuing innovative research in this area (see FAA 2010). However, the evidence is so far very much focused on delays to aircraft rather than to passengers. The two may differ because passenger missed connections may mean that delays to passengers may be longer than delays to aircraft. Since this book is focused on costs to passengers, and investment analysts would find evidence on costs to users hard to come by, the issue is ignored from the project examples. However, as evidence on passenger delays emerges in the future, it is a cost item that is bound to become conventional on economic appraisals in aviation. See footnote 8 on section 1 of Chapter 4 for a tentative illustration.

C and 'Alternative', where the latter describes the generalised cost to passengers (for which the project airport is the preferred choice) of diverting to the alternative means of travel. Generally, this would consist of an alternative airport from which to access air travel, but for shorter routes it may also mean alternative land modes of inter-city transport, such as rail. At that point, the generalised cost experienced by the average traveller would be g_1.

Returning to the rationing scenario, as demand grows beyond D_0, there is a discontinuous one-off jump of generalised cost between g' and g_1 caused by rationing. This would imply that there is some traffic that would be willing to travel through the airport if there was capacity available, but for which the cost of diverting to the alternative means of travel, at g_1, is too high. Such potential traffic will therefore not travel, and consists of *deterred* traffic. By the time demand conditions are as described by schedule D_1, such deterred traffic is measured by the difference between q_d and q'.

As demand continues to grow, once D intersects the 'Alternative' schedule to the right of R, there will be traffic that uses the alternative travel means at a generalised cost of g_1, even though it would have preferred to travel through the airport, at generalised cost g_0. This traffic is called *diverted* traffic. By the time demand grows to D_2, traffic at the local airport would be q', but there would also be substantial local traffic diverted to the alternative travel means (q_c-q') and deterred traffic (q_e-q_c); the latter is also called *generated* traffic.[3]

A project to expand total airport capacity beyond q' would produce benefits to three categories of passengers. First, it would save *existing* traffic (passengers that would use the airport both with and without the project) the cost of congestion (g'-g_0), equivalent to area g'efg_0. Second, it would save diverted traffic (q_c-q') the additional generalised cost incurred in using the alternative means of travel (g_1-g_0), equal to the area dacf.

---

3 Traffic (q_e-q_c) is deterred in the sense that it is deterred from travelling by the absence of the project, and it is generated in that it travels only because of the project. The terms 'deterred' and 'generated' can be used interchangeably.

And third, it would accommodate generated traffic (q_e-q_c). Such generated or deterred traffic would include passengers ranging from those who were just about to accept that they would incur g_1 to those who just about accept g_0; the latter category of passenger is called the 'marginal traveller'. Demand schedule D_2 depicts the declining reservation price (i.e. the maximum willingness to pay) of each subsequent passenger. The welfare gain to these passengers from the project would be equal to area abc, measured by the expression $((q_e-q_c) \times (g_1-g_0))/2$, equal to half the benefits per passenger that would have accrued to the same amount of diverted passengers. The division by 2 is an approximation to the actual welfare gain, which would ultimately depend on whether the demand curve between points a and b in Figure 3.1 is actually a straight line. Such an approximation is called the 'rule of a half'.[4]

---

4   Formally, by 'rule of a half' is meant the whole trapezoid area dabf, even though it is only generated traffic that is divided by half. Presumably the reason is that the formula for the area of a trapezoid used to measure the welfare gain jointly to existing, diverted and generated traffic uses a division by a half. A related issue is whether welfare gains to diverted traffic should be divided by 2 or not, which will depend on the extent to which diverted traffic can be considered as existing traffic. It is common in the literature to find that diverted traffic is also divided by half (see World Bank 2005). However, that is generally justified where diversion occurs because of a lowering in the relative generalised cost between the transport modes between which diversion occurs, but not when it is a result of removing a capacity constraint. In the current context, the term 'diverted' is used for air transport traffic that uses alternative airports as a less preferred choice because of lack of capacity in the project airport. For short-haul routes, such diverted traffic may also use modes that are available to the user at a generalised cost no greater than that of conducting the trip through an alternative airport. Diverted traffic that results from a constraint in capacity would have used the airport just as existing traffic does, had there been sufficient capacity available. There is no reason to divide by half the surplus of such traffic, since their gain in welfare from the project is as high as the welfare gain for existing traffic. In practice, since such distinctions among traffic categories are generally very demanding in terms of passenger data, a pragmatic alternative would be to treat diverted traffic as homogeneous by assuming that diverted traffic shares common characteristics besides an average value of time, such as location relative to the two airports, implying that marginal and average cost of diversion are equal for that traffic category, removing the need to divide by 2 the welfare gain to diverted traffic.

As is mentioned at the beginning of this section this same model can be applied to the case of an airline. If the airline has a monopoly on a route, the analysis can be replicated conceptualising the airline in place of the airport. The 'Alternative' would then represent either an alternative airline offering the same city pair but with an intermediate connection, or the road, rail or sea transport modes. If instead the airline competes with other airlines offering also direct services on the route, the alternative would become other airlines offering alternative, less convenient departure times, or fewer departure frequencies. If the competing airlines offer schedules of comparable quality, the generalised costs become very close, products are less differentiated and the situation becomes close to perfect competition.

In practice, the analysis depicted in Figure 3.1 can be usefully applied to estimate project returns when there is sufficient permanent differentiation in product attributes between the alternatives, with corresponding differences in generalised costs. This usually involves airports competing with other airports or other transport modes, or air navigation service providers (ANSPs) serving airlines that can choose different routes with alternative ATM suppliers, which may happen mostly on long-haul trips. When the competitors offer similar products, so that the generalised costs offered to users do not differ much among competitors, the outcome approaches perfect competition, in which case, as discussed in Chapter 2, section 7.1, the benefits of the project would be the financial returns after correcting for any price distortion. The corollary of this discussion is that the difference between g_0 and g_1 measures the degree of competitive advantage (in terms of granting the customer additional value) granted by the project to the promoter, in a way that promoter revenues cannot measure. Measures of competitive advantage are illustrated in project examples analysed later in the book.[5]

Likewise, the analysis is valid for freight transport. For most freight categories, however, the room for product differentiation

---

5   For Airports see Chapter 4, section 2; and for airlines see Chapter 6, section 3.

through generalised cost is somewhat narrower than for passengers, especially in terms of choice of departure and arrival airports. This is because users may have lower values of time and be less sensitive to departure time and in-vehicle time, although much difference could be expected across product categories.

In terms of inter-modal competition in freight, air transport as a whole can still develop definitive competitive advantages. The more perishable the good the higher the willingness to pay for time savings and, hence, the greater the responsiveness to time differences. In extreme cases, some industries such as year-round intercontinental delivery of fresh flowers can only be viable through air transport. In such cases, the absence of air transport would imply that deterred traffic would consist of the local flower export business as a whole. The benefits to the local economy can be substantial (more on this in section 3.3 below).

2   Airside Investments

Airside investments aim at increasing the number of aircraft movements or the size of aircraft a system can process. These outcomes constitute two sources of benefits. First, an increase in the capacity to handle aircraft movements implies an increase in departure frequency. This has the effect of reducing the frequency delay – or the time the average passenger or freight shipper has to wait until the next departing flight – and hence the behavioural or user's generalised cost of transport. Relevant investments include building a new runway or taxiway in an airport, or increasing the capacity of ATM through, say, investing in ATM equipment to enable reduced vertical separation.

The second source of benefit arises from enabling the operation of larger aircraft, which brings about improvements in operating costs because larger aircraft are cheaper to operate on a per-seat basis. These types of investments would apply exclusively to airports. There is no ATM equivalent to this second benefit, as smaller, propeller aircraft tend to be slower, requiring more ATM capacity. That is, there is an inverse relationship between aircraft size and ATM capacity requirements, and a direct relationship between aircraft size and airport capacity requirements. Still, ANSPs can influence aircraft size by constraining airspace flight movement capacity.

However, there is often a trade-off between the two sources of benefit. Airlines, for example, when replacing or expanding capacity weigh the extent to which the new capacity should take the form of more aircraft or larger aircraft. Emphasising more aircraft would enable greater departure frequency and more direct destinations to be offered, improving the quality of the airline's schedule; whereas emphasising larger aircraft creates the potential for cheaper tickets; it may also have some comfort advantages.[6]

The decision of airlines and airports are not independent. In deciding on fleet composition, airlines need to consider constraints on airport capacity. For example, constraints on the availability of slots at their hub airports mean that airlines are forced to tilt their decision towards greater aircraft size, rather than greater departure frequency. Similarly, airports that expand capacity tend to take into account the capacity requirements of the fleets of the main airlines serving the airport, which may require adjusting terminal, apron, taxiway and runway sizes.

Considering airlines and airports together, the trade-off between aircraft size and departure frequency is depicted in Figure 3.2. The downward-sloping FD curve represents the marginal frequency delay, which decreases with flight frequency. The monetary value of the frequency delay is measured along the left vertical axis. For a given number of seats supplied, frequency delay varies directly with average aircraft size (AS) – that is, for a given number of seats, the larger the aircraft size the lower the departure frequency and the higher the frequency delay. Therefore, the FD schedule increases (decreases inversely) with aircraft size, as depicted on the right vertical axis.

---

[6] There is some evidence that passengers attach a comfort value to larger aircraft. See, for example, Coldren et al. 2003 and Ghobrial 1993. However, as is argued in Chapter 2, section 6, comfort issues are ignored.

The Basic Framework 49

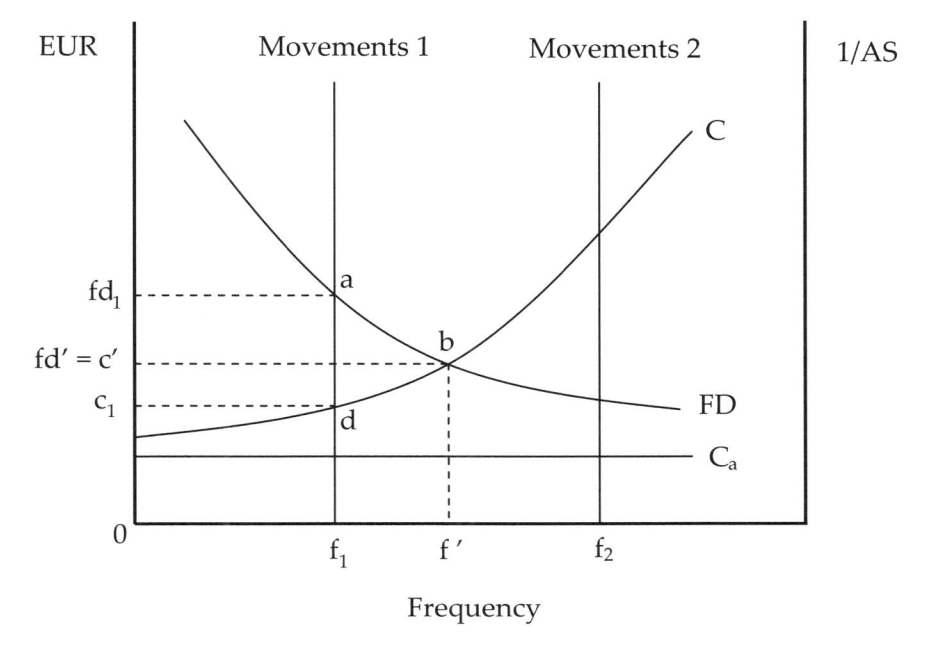

Figure 3.2   Costs in airside capacity provision

The horizontal Ca schedule represents the marginal cost to the airport of adding an extra flight, assuming constant returns to scale to provision of airside capacity. The C curve represents the total cost, including both airport and aircraft costs. It is upward-sloping because, for a given number of seats, as frequency increases the average size of aircraft decreases, increasing per seat costs since smaller aircraft have higher unit costs.

The vertical Movements 1 and Movements 2 schedules represent frequency capacity of the system before and after airside expansion, respectively. The Movements 1 schedule can be thought of as the departure frequency capacity of the airport with only one runway, equal to f_1, and Movement 2 as the frequency capacity adding a second runway, higher at f_2. The 'Movement' schedules can also represent two airspace capacity levels, before and after equipment enhancement.

When airside capacity is at Movements 1 and departure frequency is constrained at f_1, the marginal benefit of adding a departure frequency is fd_1 on the left vertical axis. This is higher than the marginal cost of decreasing aircraft size, equal to c_1. By expanding airside capacity to Movement 2, flight frequency would increase to f', which would be accompanied by a decrease in aircraft size. At that point the marginal benefit of improving frequency delay is equal to the marginal cost of decreasing aircraft size ($fd'=c'$). The benefit of expanding airside capacity from Movement 1 to Movement 2 would be equal to the area abd.[7]

3   Scenario-building

Investment appraisals aim at measuring what producers and – when the appraisal is economic – consumers and society at large gain as a result of an investment, relative to what could be expected to happen should the investment not take place. That is, project benefits and costs are measured in incremental terms. Investment appraisal therefore relies on building at least two scenarios. First, the project, or 'with project' scenario, describing what is expected to happen regarding key input and output variables during the implementation and operation of the project. And second, the counterfactual or 'without project' scenario, including assumptions about what could be expected to happen should the project not be carried out. The degree of competition in the market where the investment project takes place plays a central role in building the scenarios. A high degree of competition would imply that competitors would tend to offer similar products to those offered by the project promoter, restricting the options of the promoter in the 'without project' scenario. If instead, competition is feeble or practically non-existent, the analyst would have to make assumptions about the 'without project' scenario in two respects: the behavior of the promoter; and the amount of available capacity in the market should the project not be carried out.

---

[7] Note that by building the second runway there would be – at least initially – excess runway capacity. While this may well be the welfare maximising option, generally traffic growth means that capacity is eventually filled. Supplying facilities that operate at less than full capacity stems from technological indivisibilities in production functions.

The Basic Framework

This section of the chapter addresses these issues in turn. Beforehand, however, it deals with another issue that is the subject of much variation in scenarios built up in project appraisal in practice, namely: whose benefits and costs are accounted for? Are they those of the world at large, the nationals of a particular country, or local residents? The results of an investment appraisal will differ if particular groups are not accounted for, or if the benefits of some groups are given greater weight.

3.1 Whose Benefits and Costs?

Financial appraisals include promoter income and costs and consider all users regardless of their provenance. Economic appraisals should include benefits and costs from all users and non-users affected by a project, including competitors, regardless of their provenance. The analysis then answers the question whether the project constitutes an efficient allocation of resources and, therefore, whether the world would be better off with the project.

Sometimes economic appraisals are conducted paying attention to who benefits, who pays and who loses. Decision-makers may have distributional objectives, may be concerned with benefits to locals or nationals, or may be particularly interested in revenues from non-residents. There is not necessarily anything methodologically wrong with such appraisals, so long as the analyst and the decision-maker are aware that the appraisal is more concerned with distributional issues than with economic efficiency. In addition, when such distinctions among stakeholders are made, building scenarios that totally exclude groups who are deemed not relevant may create confusion in the calculation process by, for example, making the measurement of capacity utilisation more difficult. It is generally a better approach to consider all stakeholders in the estimation process and then attribute different weights to the benefits and costs of different groups.

The analyst should be aware that by disregarding costs and benefits to specific groups, the appraisal exercise runs a risk of reaching counterproductive outcomes. In particular, aviation projects enjoy cost economies through capacity utilisation (economies of density), vehicle or facility size (economies of scale), and joint service to different traffic categories, such as passenger and freight (economies of scope), as well as network benefits to passengers (range of departure frequency and destinations in hub-and-spoke networks). In such contexts passengers exert positive externalities on each other and it is erroneous to assume that subtracting the benefits of one traffic category leaves the benefits to other categories unchanged. For example, an air route from A to B may enjoy a departure frequency of four flights a day with low costs per seat through use of larger aircraft only because traffic density is increased by connecting passengers from other destinations. Without those connecting passengers the route may sustain a lower departure frequency and higher unit costs (through use of smaller aircraft), reducing the welfare of origin–destination users from A and B.

This book follows a 'world economy' viewpoint in the construction of scenarios, ignoring issues of provenance or distribution, as is done in traditional financial appraisals, and following the standard tenets of welfare economics.

3.2 Degree of Competition

The need to build an ad hoc counterfactual scenario in investment appraisal is governed by the competitive conditions that characterise the market where the investment project takes place. The degree of competition ranges from a perfectly competitive market, where the number of competitors is or can be very large, to a natural monopoly, where there can only be a single viable supplier. In between these two extremes there is a continuum array of possibilities of market conditions, for which industrial organisation (or industrial economics) offers a number of generic models, such as duopoly, oligopoly and monopolistic competition, built from premises about issues such as barriers to entry, cost economies, synergies and product differentiation, among others.[8]

---

8   See any textbook on industrial organisation, for example Belleflamme and Peitz 2010 or Martin 2010.

It is very rare to find either perfectly competitive markets, where producers readily substitute each other at no cost to the consumer or to society, or natural monopolies that have no substitutes at all. Moreover, in practice there is always some degree of product differentiation, if only because of brand image. For investment appraisal purposes, the issue is one of judging the degree of competition in the market where the investment project takes place. The key judgement to make is whether, in the absence of the project, there are other firms (existing or potential entrants) that would be able to supply the market at the same or very similar conditions as the promoter. If the answer is yes, the project can be considered to be carried out in competitive markets. In that case, there is no need to build an ad hoc counterfactual, since in the absence of the project, some competitor would supply the consumers otherwise supplied by the promoter, and do so at the same, or very similar, quality and price. In effect, the counterfactual is simply the opportunity cost of the resources invested in the project.

If instead the answer is no, then in the absence of the project the consumer is dependent on the conduct of the promoter. The consumer has either no alternative supplier, or would experience switching costs to access the closest substitute, involving a loss of welfare. In such a case the analyst must make assumptions about what supply conditions the market would face should the promoter not carry out the project, meaning that the analyst must design an ad hoc counterfactual scenario. Building a counterfactual scenario would involve two critical dimensions: the actions assumed by the promoter should the project not be carried out; and the assumed capacity available in the market should the project not be carried out. These two issues are treated in turn in the next two sections of this chapter: 3.3 and 3.4.

Table 3.1 summarises the generic competitive situations that the analyst is likely to find when appraising aviation projects. Whereas the table is self explanatory, three issues may merit further explanation. First, note that it does not really matter whether the underlying competitive structure resembles more a perfectly competitive market or an oligopoly. In either case, should the project not be carried out, the market will be supplied by another competitor. In the case of perfect competition this would occur either through established players or through new entrants. In the case of oligopoly it would be by existing players expanding production.

The second issue follows from the first. The fact that under sufficiently competitive conditions substitutes are available in the primary market has implications also for the impact of the project on secondary markets. If it is assumed that without the project the same or a similar product would be supplied anyway, no effect on secondary markets can be unambiguously attributed to the project. As will be seen, this is particularly relevant for the aircraft manufacturing sector, which tends to be competitive. There, the output – namely, the aircraft – is operated in a secondary market, which may be subject to distortions, such as externalities. It is legitimate to attribute such externalities in the secondary market to a project (which takes place in the primary market) if the project affects perceivably the conditions in the primary market, which in turn affects the conditions in the secondary market, leading to more externalities. For example, if an aircraft manufacturing project affects prices in the primary market, and hence affects the total number of aircraft sold, there will be more aircraft in operation and more external costs in the secondary market as a direct consequence of the project. But if the primary market is sufficiently competitive, so that in the absence of the project other aircraft makers would take up the production otherwise carried out by the project promoter, no changes can be attributed to the project in the secondary market, including external costs.

Note that the issue refers to whether the project distinctively affects output (or prices) in the secondary market, an issue discussed in Chapter 2, section 7.3 This is a related issue to that of adjusting prices from secondary markets affecting directly the primary market, such as taxes on inputs to the project, discussed in Chapter 2, section 3.2.

The Basic Framework

Table 3.1 Competitive conditions and scenario-building in investment appraisal

Degree of competition	Key characteristics	Instances in aviation	Treatment in investment appraisal
Perfect competition	• Many competitors • Many potential entrants and no barriers to entry • No product differentiation and no pricing power by any firm • Profitability kept at opportunity cost of capital	• Airlines in dense route between uncongested airports • Inter-hub competition on intercontinental markets • Suppliers of standardised components in aeronautics	• No need to define ad hoc counterfactual • Secondary markets remain unaffected by project • Lack of project has no wider consequences
Oligopoly	• Few participating firms • Little or no product differentiation • High entry barriers but incumbents keep each other in check • Higher profitability than in perfectly competitive market	• Large airlines competing within a hub • Manufacturers of 'industry workhorse' aircraft models • Airports on a multi-airport city	• No need to define ad hoc counterfactual • Secondary markets remain unaffected by project • Lack of project has no wider consequences
Monopolistic competition	• Differentiated product gives company market power in market segment • Entry possible but may be costly • Company enjoys monopoly rents while there is no entry, although demand curve reflects competition from differentiated products	• Airline offering 'low-cost' services from a distant airport • Aircraft manufacturer on niche aircraft segment • Airports in nearby cities with overlapping catchment areas • ANSPs on alternative routes	• Need to define ad hoc counterfactual • Secondary markets may be affected by project • Continued lack of project may involve non-extreme wider consequences
Monopoly	• Only one firm in the market • Competitive entry de facto impossible • The firm sets price and quantity to maximise profit, unless subject to government regulation	• Sole airport on remote island • Sole air service to remote location • Approach or domestic ANSP • ANSP serving large areas of oceanic airspace	• Need to define ad hoc counterfactual • Secondary markets may be affected by project • Continued lack of project may involve extreme assumptions

Third, where the promoter has market power – meaning monopoly situations and, to a lesser extent, under monopolistic competition – refraining from carrying out investments may involve serious consequences to the local economy. For example, preventing a remote location from accommodating growing demand for air transport services by denying it additional airport capacity may disrupt the economic development of the region. This issue is discussed below in section 3.4, but it is worth highlighting at this stage that such scenarios are sometimes erroneously used in situations where there is de facto competition, even if the facility at hand, say the local airport, may be perceived as a local monopoly. Many subsidised airports are better closed down than expanded (with subsidies) if there is adequate surface transport to alternative airports. Shutting down the airport may actually help the local economy by saving it unnecessary subsidies.

The same logic underlies the often used (and erroneous) arguments of impacts on the local economy as a justification for building local airports. If sufficiently good services are available to airports in nearby cities, there is a good chance that the local airport (despite its apparent local monopoly position) will constitute a wasteful investment. Many of the benefits registered by the project would constitute transfers from the alternative, nearby facility. The key to the problem is recognising that not all sole local suppliers enjoy monopoly power, but may be instead taking part in monopolistic competition.[9]

3.3 Counterfactual Behaviour by the Promoter

The previous section of this chapter showed how a sufficient degree of competition does away with the need to define an ad hoc counterfactual scenario describing what would happen in the market if the promoter did not carry out the project. Where competition exists, so long as the project is sufficiently profitable, competitors will carry out the project if the promoter fails to do so. Where competition is sufficiently imperfect to

---

[9] The conceptual tools of industrial organisation used for defining markets, which are central to the practice for competition policy, may be useful to the analyst in understanding competitive conditions in the market where the investment takes place. See Belleflamme and Peitz 2010, Motta 2004 or, for a very accessible guide, Fishwick 1993. Marketing references also offer valuable insights. See in particular Lambin et al. 2007.

grant the promoter a large degree of market power, promoter behaviour in the absence of the project is not forthcoming. Defining an ad hoc counterfactual scenario is necessary in order to compare the project with what could be expected to occur without it. There are three basic types of counterfactual scenarios regarding promoter behaviour, as follows:

1. **Do nothing**: This assumes that the counterfactual to the project is that no investment takes place at all and, hence, that the capacity will gradually deteriorate, reducing the future ability of the facility to accommodate traffic. This type of 'without project' scenario is suitable for projects that consist of facility rehabilitation.
2. **Do minimum**: The 'without project' scenario assumes that there will be sufficient investment to keep the existing capacity operational. It is a suitable counterfactual for capacity expansion projects. The investment analysis would compare the project against making the necessary investments to keep installed capacity operational for the full life of the project.
3. **Do something (else)**: The 'with project' scenario is already a 'do something' scenario. A 'do something (else)' scenario would consist of an alternative approach to meet the objectives of the project. It is therefore an appropriate counterfactual for analysing project options, once it has been recognised that 'something' must be done. For example, an airport might expand capacity by building a second terminal or by expanding an existing terminal. A cargo airline might replace an ageing fleet of freighters by buying new freighter aircraft or by converting passenger aircraft into freighters.

A common source of error in scenario-building involves mixing counterfactuals 1 and 2. This might happen when a management team confronts the question 'do we expand capacity?' and then carries out the investment analysis by comparing the project against a 'do nothing' scenario, instead of a 'do minimum'. By setting 'do nothing' as the counterfactual to the project, the question that management is really asking is 'do we expand the airport or do we let it slowly degrade?' which is not the same as 'do we expand the airport or keep capacity at current levels?' If what management mean to ask is the latter question, but they define the analysis through the former question, they will tend to overestimate the returns of the capacity expansion, which may lead them to take a wrong decision, probably by overinvesting.

The third type of counterfactual refers to the opportunity cost of the project. Depending on what the remit of the analyst is, it may not be enough to compare a project against a 'do nothing' or 'do minimum'. The analyst may be asked to check whether there are better project alternatives that would maximise value for the company or for society. In competitive situations, not following the best alternative opens the way for a competitor to adopt it and develop a competitive advantage.

3.4   Counterfactual Capacity

Passenger behaviour in the 'without project' scenario will be determined by how much alternative capacity is available and under what conditions. When markets are competitive, the answer is straightforward: in the absence of the project, competitors would supply a similar amount of capacity, and at similar price and quality conditions, to what the promoter would supply with the project. However, when competition is poor, the amount of capacity available in the market should the project not take place may not be obvious. And yet, knowing the counterfactual capacity is necessary for the analyst to estimate diverted and generated traffic. This gives raise to two potential problems. The first is that it is not always possible for the analyst to be certain about available capacity without the project. When that is the case it is quite likely that the analyst will have to contemplate the possibility that, at some point in the project life, any counterfactual capacity would entail much poorer generalised cost conditions. This leads to the second problem, which is that even when the capacity conditions in the alternative scenario are known but are much inferior to those supplied by the project, the analyst may be forced to make extreme assumptions in the 'without project' scenario, which would involve difficult to quantify knock-on effects on the local economy.

To illustrate the discussion that follows, let us return to Figure 3.1 above and assume that it consists of an airport project. Generated traffic is q_e-q_c, and diverted traffic q_c-q'. Say that the difference in generalised cost between the 'with project' and 'without project' scenarios consists of two hours of traveller time for the average passenger. That is, the difference between g_1 and g_0 on the vertical axis of Figure 3.1 is accounted for by two hours worth of passenger time alone. In the case of an airport, those two hours can refer to the additional time incurred by driving to the alternative airport, or to the average delay to an alternative, less convenient departure time at the project airport. To simplify, it is also assumed that this is the case per passenger for the entire life of the project, which is why the schedules relating generalised cost g to traffic are horizontal.

Such a scenario carries with it an implicit assumption, which becomes increasingly artificial as one looks further into the future within the project lifespan. The implicit assumption is that there is sufficient existing capacity in the airport where the project will take place (the project airport), and/or in the alternative airport, to accommodate all diverted traffic throughout the life of the project. In reality, this may be so only in very particular circumstances involving substantial overcapacity during off-peak periods at the project airport and/or in the alternative airport. Traffic diverted to the alternative airport in the 'without project' scenario will use up capacity at the alternative airport that was originally planned for traffic in the more immediate catchment area of that airport. At some point in the future, the growing traffic diverted from the project airport to the alternative airport will bring forward in time any need for capacity increase at the alternative airport. Therefore, a realistic 'without project' scenario may involve assuming capital investments in the alternative airport a few years into the project life. This brings us into the conundrum that the 'without project' scenario may involve an investment equivalent to that in the 'with project' scenario at the project airport, but at an alternative airport and possibly further into the future.

The conundrum is all the more puzzling for traffic which in the 'without project' scenario is diverted to inferior off-peak flight times within the project airport. At one point there will be no capacity left at inferior times. The 'without project' scenario would reasonably have to include investment in additional capacity. That is, the alternative to the project would be the project itself. Alternatively the 'without project' scenario would soon imply extravagant assumptions.

To see this, let us consider the case where there is no alternative airport, and where the existing airport has substantial market power. A hypothetical extreme case would be a remote island with a single, highly congested airport. Here, the alternative to air transport would be much inferior – say an eight-hour flight would have to be substituted by a week-long ship voyage. The delay experienced by diverted traffic (to inconvenient flight schedules) will grow longer over time as the airport faces growing demand and congestion. As a result, a growing share of traffic in the 'with project' scenario will constitute traffic generated by the project, rather than diverted traffic. This is illustrated in Figure 3.3, which restates Figure 3.1, leaving the now non-applicable 'alternative' generalised cost schedule in the background as an intermittent dash-and-dot line. Figure 3.3 also introduces an alternative exponential cost line C_{exp}, depicting exponentially growing delays as the airport gets increasingly congested even during less preferred travelling times (off-peak would no longer be a valid description as the airport will tend to become equally busy throughout the day). Finally, to improve clarity Figure 3.3 also removes demand curve D_1.

In Figure 3.1 traffic generated with the project (or deterred without the project) was q_e-q_c, and diverted traffic q_c-q'. Now, in Figure 3.3, assuming that the generalised cost experienced by travellers relates to traffic as depicted by schedule C, generated traffic grows its share of project traffic to q_e-q_{c2}. In the 'without project' scenario, if rationing is implemented as depicted in

The Basic Framework 61

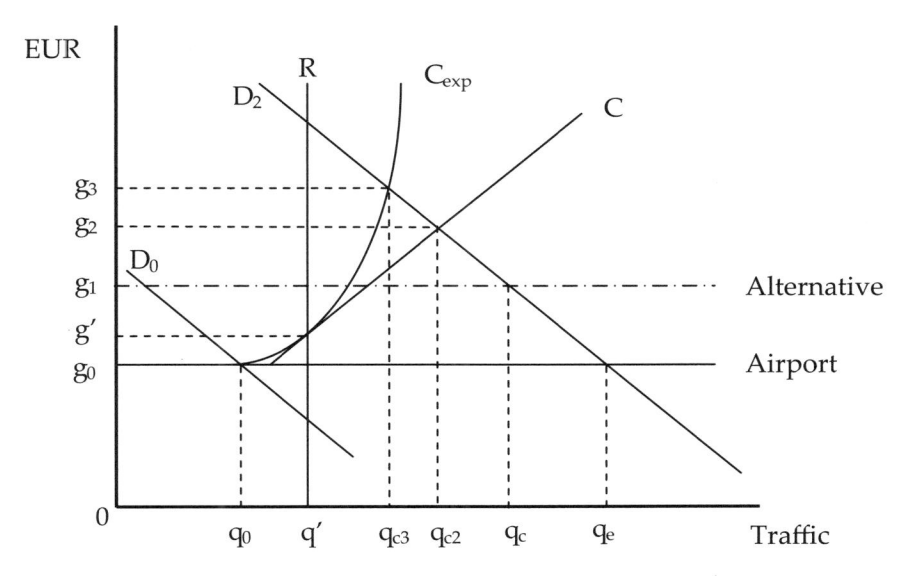

Figure 3.3 Alternative counterfactual capacity conditions in the presence of market power

schedule R, an amount of traffic equal to q′ could be expected to travel at reasonable times, experiencing a generalised cost of g′, and q_{c2}-q′ traffic will experience diversion to less than preferred departure times, causing their generalised cost to increase to g_2. Perhaps more realistically, the schedule delay and associated diversion costs will grow exponentially as depicted by schedule C_{exp}, with traffic experiencing increasingly higher costs as available flight times are pushed into more inconvenient times. Deterred (or generated) traffic would grow to q_e-q_{c3}, and diverted traffic would diminish to q_{c3}-q′. Soon, thereafter, as C_{exp} tends towards verticality, additional demand for air travel will be deterred from travelling altogether and the costs imposed on travellers will be ever-growing.[10]

---

10 In such a case, when computing total user costs, making a distinction between traffic experiencing a generalised cost of g′ and traffic experiencing a generalised cost of g_3, is largely circumstantial, depending on scheduling practices and how the 'without project' scenario is defined. An alternative way to frame the scenario would be to view the C or C_{exp} schedules not as marginal delay but as average delay schedules, in which case the distinction

An ever-rising generalised cost g up the vertical axis in the 'without project' scenario will imply extravagant assumptions in the investment appraisal exercise. Denying an area access to highly demanded airport capacity when alternative transport means are much poorer – as tends to be the case in medium- to long-haul air passenger transport – would eventually start preventing local firms from generating economies of scale and prompt industry relocation. The 'with project' and 'without project' scenarios would have to assume differential knock-on effects on the local economy, affecting productivity and income levels.[11] The costs of the 'without project' scenario become very large, dwarfing any costs of carrying out the project, leaving virtually any airport project, however costly, worthwhile. Quantification of benefits and costs both to users and to the local economy becomes difficult, rendering the exercise highly speculative.

The analyst faces two possibilities then. One is trying to estimate the dislocation caused to the local economy. The exercise is complex because of the many sectors involved and the computational difficulty of attempting to estimate the effects on each of them. In any case, a key point to take note of is that, in such an exercise, most of the traffic growth with the project will constitute generated traffic, and any producer surplus attributable to that traffic will constitute a net project benefit.

The second possibility for the analyst is to recognise that the 'without project' scenario involves also investing in capacity expansion, on top of maintenance costs, to keep existing

---

disappears. After all, should such a scenario materialise, airlines lucky enough to have slots at the most desirable hours of the day will tend to respond to demand pressure by increasing the price of air tickets for flights at those preferred hours, appropriating any consumer surplus available. In that case, the increased profitability of the airlines will constitute a welfare transfer from the passengers to the airlines.

11   In graphical terms, the situation could be illustrated with the analysis in Chapter 2, section 7.3. The demand curve faced by multiple secondary markets in the local economy in the 'without project' scenario would fall to the left of where it would fall in the 'with project' scenario, implying higher prices on multiple goods and services to the local population.

capacity operational. This has implications for the treatment of producer surplus, which are discussed next.

3.5   Producer Surplus

Before picking up the discussion begun above, it is important to clarify two preliminary issues regarding producer surplus. The first is that when dealing with the producer surplus of a project, the analysis does not measure promoter profitability as a whole, but the incremental profits that result from the project. That is, the difference between the profits that the operator makes with the project and the profits that the operator could be expected to make in the absence of the project. So it is possible for an airline, say, to make a bad investment in aircraft, one that generates losses, while the airline remains profitable. The link between the appraisal of the investment project and the wider profitability of the airline is that, other things being equal, the negative returns shown in the investment appraisal exercise will bring about a lower degree of profitability to the airline as a whole, following the implementation of the project, relative to the profitability the airline would have achieved had the project not been carried out.

The second preliminary issue regarding producer surplus relates only to the economic profitability of the project, not to the financial profitability of the promoter. When looking at society at large, the incremental profits to take into account are both those of the project and those of the alternative service that users would have used had the project not taken place, regardless of whether that alternative service were run by the promoter or by an alternative operator. So, if one investment project simply switches traffic from one service to another (diverted traffic) and the profits of both services are equal, total producer surplus, as far as society is concerned, remains unchanged. On an economic appraisal, only profits arising from generated (or deterred) traffic, that is, traffic that would not travel at all should the investment project not take place, amount to a net gain in producer surplus, and hence in social welfare. Profit from diverted traffic would cancel out.

The discussion in the previous section concluded that in situations when the project alternative is much inferior to the project, so that not carrying out the project may imply serious damage to the local economy, dwarfing any project cost, there are two possibilities. The first is to attempt to estimate those costs to the local economy, a task of great complexity. In such a scenario most of the traffic would be generated traffic and hence (most, or for simplicity, all of) any producer surplus could be considered a net benefit of the project. The second possibility is to recognise that the 'without project' scenario also involves investing in capacity.

Should the analyst follow the second alternative, the exercise then becomes akin to that of investing in marginal capacity in a competitive market, whereby if one operator does not do it, some other will. In practical terms, the absence of the project would involve confronting the market with a vertical supply curve. The role of the project would then be to contribute to ensure that the supply curve continues to be horizontal. In that case, standard economic appraisal would recognise that the economic return of the investment project corresponds to the marginal return of an increase in capacity, which corresponds to the financial return of the project, or the growth in producer surplus.[12] As with the first possibility, discussed in the previous paragraph, the full producer surplus yielded by the project (without making a distinction between the portions attributable to diverted and to generated traffic) is taken as a social benefit.

3.6   Conclusion

When carrying out an investment appraisal, the analyst should first identify the competitive nature of the market concerned. The key assumption to make is whether in the absence of the project there are other firms that would be able to supply the market at similar conditions as the promoter. If yes, there is no need to build an ad hoc counterfactual scenario.

---

12   As in all economic appraisals, ensuring that changes in producer surplus allow for any transfers in the form of taxes, subsidies or social security contributions.

The counterfactual is simply the opportunity cost of the resources invested in the project. If instead the promoter enjoys a degree of market power, the alternative to the project may involve costs to the consumer and the supply conditions must be defined. The counterfactual scenario would have two dimensions: the actions of the promoter without the project, and the capacity conditions available in the market to the user without the project, by the promoter or by other operators. Regarding the counterfactual line of action of the promoter, the analyst must see that the type of action chosen, whether 'do nothing', 'do minimum' or 'do something (else)' matches the purpose of the project and, hence, the question that the analysis is aiming to answer. Each of those types of counterfactual is suitable for rehabilitation, capacity expansion and evaluation of alternatives, respectively.

Regarding the capacity conditions available in the market without the project, besides defining the amount of capacity per se, the scenario assumed also has implications for the treatment of producer surplus. The suggested way to proceed is as follows:

- When it is realistic to assume that there is sufficient alternative capacity through the life of the project to accommodate diverted traffic, the appraisal exercise should be based on estimated differences in generalised cost of travel. Of any producer surplus produced by the project, only that part attributable to *generated* traffic unambiguously constitutes a net benefit to society. Any producer surplus attributable to *diverted* traffic must be set against any producer surplus in the 'without project' scenario.
- When the available alternative is much inferior to the project in terms of generalised cost of travel (say many hours or days travelling by surface transport), the analyst should proceed with the standard process of estimating differences in generalised cost, but should bear in mind that the 'without project' scenario would ideally incorporate hard-to-quantify knock-on effects on the local economy.

Since such estimates are generally outside the scope of normal economic appraisals, the analysis will tend to underestimate project benefits. As for producer surplus, in those cases, most of any such surplus will be attributable to generated traffic and, hence, most (or, for simplicity, all) producer surplus gained with the project constitutes a net welfare gain to society as well as to the promoter.

- The tricky situation, one which nonetheless is not that unusual, is when it is not clear whether there is sufficient alternative capacity available to the project throughout the project life and, hence, whether the additional marginal generalised cost assumed in the 'without project' scenario can realistically be held constant through the estimation period. In those situations it would not be clear whether it would be more realistic to assume that marginal generalised cost in the 'without project' scenario increases substantially – probably exponentially – a few years into the life of the project (possibly including difficult-to-quantify, knock-on effects on the productivity of the local economy). It may also be unclear whether the analyst should also assume capacity investment in the 'without project' scenario. As has been suggested in this chapter, a pragmatic approach would be for the analyst to assume the generalised cost of the alternative to be as observed at the time when the investment appraisal exercise is carried out, but then to consider the whole of the incremental producer surplus yielded by the investment to the promoter (without making a distinction between diverted and generated traffic) as a net gain to society. This would be in recognition that the project will very likely constitute adding necessary marginal capacity, akin to the treatment of producer surplus in a competitive market situation, while acknowledging some degree of product differentiation as determined by the differences in generalised cost between the project and the alternative.[13]

13　Again, this is a pragmatic, operative suggestion for appraisals in practice. It does not claim to be accurate about the estimates of investment

Such an approach recognises, after all, that the project is increasing the total productive capacity of the transport system.[14]

One final additional factor to bear in mind when dealing with producer surplus is that the analyst must be careful to establish that the producer surplus results from added capacity, not from increased prices. Price changes involving supernormal profits constitute welfare transfers between consumers and producers; they also generate deadweight losses. Such a situation will be illustrated in Chapter 7, section 1.2, where investments in the aeronautical sector under conditions of monopolistic competition are discussed.

---

returns, nor does it rule out the existence of more suitable solutions. The merits of the rule proposed here are twofold: first, it is simple to apply; and second, it is hard to see in what circumstances it will yield wrong advice as to the merits to proceed with a project or not, or even while ranking alternative projects.

14   Which, incidentally, would also serve as available 'without project' capacity to any investment appraisal exercise of an eventual project to build an alternative to the island airport, such as a second airport on the island.

Chapter 4
Airports

Introduction

This chapter addresses the economic appraisal of airports, putting in practice the concepts introduced in chapters 2 and 3. It also illustrates a number of economic policy issues that manifest themselves in the economics of investment. The sequence of project types begins with a greenfield airport in section 1, followed in section 3 by a terminal capacity expansion, which builds on the greenfield airport case. Airside capacity projects are illustrated through the appraisal of enlarging an existing runway in section 6 and adding a new runway in section 7.

Policy issues are introduced in self-contained sections alongside these project examples. The policy issues are not necessarily specific to each type of project. Rather, they are introduced whenever the results of a case invite discussion. Three of the policy sections concern the suitability of private sector involvement in infrastructure investment, including identifying when there is room for such involvement (sections 2 and 4); and how the regulatory framework affects the incentives the private sector operator faces in the investment decision (section 8). The latter case is linked to the generic incentive to overinvest brought about by rate of return regulation, discussed in section 5.

Whereas the focus of the case presentations is on illustrating estimation processes, and it is not the intention of the book to arrive at any policy recommendation, the numbers used in the examples are realistic. However, they do not relate to any specific real-life project.

1   A Greenfield Airport

Town A and its conurbation have experienced significant population and income growth following the discovery of mineral deposits just over a decade ago. Population is currently around 200,000, with an average income per capita of €15,000 per year. Town A has no airport and locals use the nearest airport, which is in town B, about two hours' drive away. Given the already sizeable and growing population, and the distance to town B's airport, the local authorities wondered whether they should develop a local airport and hired airport planners to estimate local demand and propose an airport project.

The planners began by estimating the potential air transport demand generation in the area, depicted in Table 4.1. They surveyed demand in the country, estimated the catchment area, built econometric models and found out that at that income level, the region should generate about 1.5 trips per inhabitant, or 300,000 trips in total per year. Local economic forecasts assumed that the population would continue to grow at a cumulative annual growth rate (CAGR) of 1.2 per cent per year; so that in 20 years' time the population would grow to 254,000 and in 30 years to 286,000.

Table 4.1   Estimating traffic-generating potential of town A

	Year	1	20	30
(1)	Population CAGR since year 0		1.2%	1.2%
(2) from (1)	Population	200,000	253,887	286,052
(3)	Income CAGR since year 0		2.0%	2.0%
(4) from (3)	GDP/capita (EUR)	15,000	22,289	27,170
(5) from (4)	Cumulative income growth		49%	81%
(6)	Income elasticity of demand		1.4	1.4
(7) from (5) and (6)	Trips/capita	1.5	2.5	3.2
(8) = (2) × (7)	Trips	300,000	639,919	916,471
(9) = (8) × 2	Passenger throughput	600,000	1,279,838	1,832,942
(10) from (9)	Throughput CAGR since year 0			3.8%

The planners expected the income elasticity of demand going forward to be 1.4, meaning that a 10 per cent growth in income would bring about a 14 per cent increase in the propensity to travel. Studies showed that income per capita was expected to grow at 2 per cent per year over the long term. This would imply that over 30 years, real incomes would grow some 81 per cent to just over €27,000. Such income growth, combined with an income elasticity of 1.4, would mean that propensity to travel would rise by 114 per cent to 3.2 trips per inhabitant per year.[1] The result of this would be that in 30 years, town A would generate almost 920,000 trips per year. Since each trip involves a departure and an arrival, the airport would see a throughput of 1.83 million passenger movements. The resulting cumulative average growth rate of traffic in the period is 3.8 per cent per year.

In parallel to the desk exercise of the traffic generation capacity, the planners undertook surveys of actual travellers and found that the number of passengers from town A using town B's airport each year was around 300,000, implying 600,000 passenger movements per year. This was the same round figure which they had estimated through the desk exercise. The planners were surprised by this result, which was higher than expected, since the observed traffic estimated by the survey would exclude deterred traffic (or traffic that would be generated by the town A airport project). Something in the demand mix of town A made it a stronger generator of traffic than expected, probably attributable to the export sector playing a larger role in the local economy than the national average. Therefore it was deemed advisable to increase the traffic estimate for the airport, adding an estimate of deterred traffic. The estimation of deterred traffic was made by comparing the private (or behavioural) generalised cost of transport with and without project, then estimating the impact of their difference

---

1 'Propensity to travel' means the trip generation capability of town A, whether as origin or destination; that is, the magnitude includes both trips carried out by town A residents and trips attracted to town A from non-residents, which depend equally on the population size and income of town A. Tourism destinations and hub airports require additional considerations that are not dealt with here.

on traffic using standard demand elasticities. The calculation is summarised in Table 4.2.

Table 4.2   Estimating deterred (or generated) traffic in town A

	Generalised cost (GC) with diversion		
(1)	Avg. flight duration	hours	1.5
(2)	Avg. one-way air ticket price	EUR	200
(3)	Passenger processing time	hours	1.5
(4)	Access/egress time	hours	2.25
(5)	Access/egress operating cost	EUR	20
(6)	VoT	EUR	15
(7) = (2) + (5)	Total money cost	EUR	220
(8) = ((1) + (3) + (4)) × (6)	Total time cost	EUR	78.75
(9)	Air safety cost	EUR	1
(10)	Access/egress safety cost	EUR	3.6
(11) = (7) + (8) + (9) + (10)	**GC with diversion**	EUR	303.35
	Net cost of diversion		
(12)	Access/egress time	hours	2
(13) = (6) × (12)	Access/egress time cost	EUR	30
(14)	Access/egress operating cost	EUR	15
(15)	Access/egress safety cost	EUR	3
(16) = (13) + (14) + (15)	Total savings	EUR	48
(17) = (11) − (16)	**GC without diversion**	EUR	255.35
(18)	GC elasticity		−1.2
(19) = ((17) / (11)) − 1	Relative change in GC		−15.8%
(20) = (18) × (19)	**Deterred/Generated traffic**		19.0%

The survey found that the average duration of flights taken by air travellers from town A through the airport at town B was 1 hour and 30 minutes (i.e. 1.5 hours). The average one-way air ticket paid was €200. Time spent at the airport at both ends of the route also averaged 1.5 hours. It took on average 2 hours and 15 minutes (i.e. 2.25 hours) to access airport B from A, at a cost of €20 per trip. Studies carried out for transport planning for the regional economy saw that the average airport user travelling on business would be willing to pay about €20 to save an hour, and the average leisure traveller about €10. Surveys also showed that air transport trip purpose in the region was 50 per cent business and 50 per cent leisure, so the average value of time was €15. The safety costs of travelling by air were estimated to be €1.[2] Diversion occurred mostly by car and local transport planning parameters of road accident rates and willingness to pay for safety resulted in a cost of safety per passenger of €3.6 per one-way trip.[3] With those parameters, the generalised cost was estimated at €303.35 per one-way trip.

With an airport in town A, access and egress time to the airport would take 15 minutes, instead of the 2 hours and 15 minutes taken to reach airport B, and would cost €5 instead of €20. The shorter distance by road would also mean that the road safety cost would fall to 60 cents. Therefore the savings to generalised cost by using airport A instead of B were estimated at €48 per one-way trip and the generalised cost of travelling through airport A would be €255.35 per one-way trip. This constitutes a saving of 15.8 per cent in generalised cost of travel, which at a demand elasticity of −1.2 would translate into an estimate of traffic currently being deterred by the lack of an airport in town A of 19 per cent of observed traffic. Therefore, should the airport in town A be opened at the time of estimation, traffic at the airport would be the 600,000 passenger trips currently diverted to B, plus 19 per cent of generated traffic, or an additional 114,000 passenger trips per year, bringing total traffic to 714,000 passengers per year.[4]

---

2 The €1 result would come by simplifying the calculation by focusing on the cost of fatalities only and not of minor or serious injuries. If the chances of dying on a commercial flight were 1 in 2 million, and the value of statistical life in the country at hand was €2 million, by multiplying both figures, the value of the risk of travelling by air would work out at €1 per passenger.
3 At this stage, the only safety cost included is the value of safety or the value of statistical life, which is determined by user willingness to pay and, hence, affects traveller decision-making. Additional accident marginal costs such as medical costs incurred by the rest of society are excluded at this stage. See HEATCO 2006.
4 Note that the estimate of generalised cost could also include a measure of frequency delay. This would increase the data requirements and would necessitate making strong assumptions about flight schedules in the future airport. In addition, if it is assumed that departure frequency conditions

Given traffic projections, the planners proposed an airport with a capacity for 1.2 million passengers per year, sufficient to accommodate both current and expected demand for the next 20 years.[5] The planners estimated that the airport would cost €280 million to build. Land was relatively expensive as the airport was to be located close to town A, some 7 kilometres away. Land expropriation would cost €100 million, bringing cost to €380 million.[6] The existing access road was deemed sufficient to handle the expected increase in road traffic. However, a small suburban community would be affected by noise, so that the investment would need to include €20 million to install double glazing in buildings. All in all, the investment would add up to €400 million.

Regional politicians were keen to have an airport. However, the cost was large and the government had other pressing needs, including a large hospital. Airport consultants had said that an airport of that size would at best be marginally profitable and that the government should expect the airport to be a net financial liability over the foreseeable future. On the other hand, they pointed out that the airport would improve

---

would be similar in both airports – at least for the most preferred destinations for citizens in A – frequency delay would cancel out. It would still affect the relative difference between generalised costs and the estimate of generated traffic. In the present example, assuming an average 1.5-hour frequency delay in both airports would mean that generated traffic would be 17.7 per cent of observed traffic, rather than 19 per cent. In practice, since the difference it makes to relative generalised costs is unlikely to be large, and is based on strong assumptions, it may be simpler to ignore frequency delay altogether in terminal capacity projects, unless the nature of the project demands otherwise. As has already been mentioned, and as is illustrated below in section 8 and in Chapter 5, section 1, frequency delay plays a critical role in airside (i.e. aircraft movement capacity) projects.
5   This could be taken to consist of capacity supplied with an IATA level of service C. See IATA's *Airport Development Reference Manual*. Available at: http://www.iata.org/publications/Pages/airportdevelopment.aspx (accessed: 31 July 2013); and de Neufville and Odini 2003.
6   For simplicity, land is treated just like any other capital asset or input into the project. However, the analyst should be aware that land raises a number of issues in economic analysis related to restrictive licensing policies, price controls, expropriation policies, forced resettlement, etc., which affect the relationship between the price paid by the project promoter and the opportunity cost of land. Since these issues are not specific to aviation, they are not dealt with further here.

the accessibility of the regional economy, decreasing costs to businesses, encouraging visits by non-residents and improving leisure options for local residents.

The government wondered whether the airport should be built and commissioned a financial and economic analysis of the investment to help it make a decision. Table 4.3 presents the results of the analysis, including all construction years (years 1 to 4), the first year of operation and selected years through the life of the airport.

The first step would consist of estimating the financial returns of the project, based on total cash consumed and generated. Revenues would come from two broad sources. First, aeronautical activities, which would include charges per passenger movement, ground handling for aircraft and passengers, and aircraft landings, parking and servicing. A comparison of charges with those of other airports in the country showed that they could add up to an average of €6 per passenger. The second source would be non-aeronautical activities, including revenues from retail activities in the terminal, car parking and renting of property on the airport site. Estimates showed that non-aeronautical activities could generate €2 per passenger net of cost of merchandise sold.

Row 1 in Table 4.3 shows the design capacity of the airport. Forecasted throughput is included in row 2. The airport would have sufficient capacity to absorb all of the traffic that town A is forecasted to generate during the first few years of operation. Traffic would reach design capacity by year 15 and by year 25 capacity would start being rationed. Rows 4 to 9 show revenue and cost calculations. The net cash flow shown in line 9 includes operating revenues minus operating costs and capital investment.

The project would not be financially viable. Discounted at 5 per cent, the yield of long-term government bonds, it would have a negative present value of €394.6 million. If the project were discounted at the capital cost of the private sector, estimated at around 8 per cent, it would be even less appealing to the private sector. Therefore the project would not be carried out by the private sector without financial assistance provided by the government.

Table 4.3 Financial and economic returns of a greenfield airport project

		Year\PV	1	2	3	4	5	10	15	20	25	29
	FINANCIAL RETURNS											
(1)	Airport passenger capacity (thousand)		0	0	0	0	1,200	1,200	1,200	1,200	1,200	1,200
	Passenger throughput											
(2)	With project (thousand)		0	0	0	0	829	999	1,203	1,450	1,596	1,596
(3)	Without project (thousand)		0	0	0	0	0	0	0	0	0	0
	Operating cash flows (after tax)											
(4)	With project (EUR m)	−40.0	0.0	0.0	0.0	0.0	−4.1	−3.9	−3.5	−2.9	−2.4	−2.4
(5)	Without project (EUR m)	0.0	0.0	0.0	0.0	0.0	0.0	0.0	0.0	0.0	0.0	0.0
(6) = (4) − (5)	Net benefit (EUR m)	−40.0	0.0	0.0	0.0	0.0	−4.1	−3.9	−3.5	−2.9	−2.4	−2.4
(7)	Capital investment (EUR m)	354.6	100.0	100.0	100.0	100.0						
(8)	Subsidy (EUR m)	0.0	0.0	0.0	0.0	0.0						
	Net cash flow to operator (EUR m)	**−394.6**	−100.0	−100.0	−100.0	−100.0	−4.1	−3.9	−3.5	−2.9	−2.4	−2.4
	Operator FRR	N/A										
(10)	Government financial flows (EUR m)	89.8	15	15	15	15	2	3	3	4	4	4
(11) = (9) + (10)	Operator + government flows (EUR m)	**−304.8**	−85	−85	−85	−85	−2	−1	0	1	1	1
	Private and Government FRR	N/A										

ECONOMIC RETURNS

Diverted passengers											
(12)	With project	(thousand)		600	623	646	671	0	0	0	0	0	109
(13)	Without project	(thousand)		600	623	646	671	697	839	1,011	1,219	1,469	1,705
Deterred passengers											
(14)	With project	(thousand)		114	118	123	127	0	0	0	0	151	324
(15)	Without project	(thousand)		114	118	123	127	132	159	192	231	279	324
Cost of diversion											
(16) = (12) × time cost	Time cost with project	(EUR m)	70.0	17.6	18.7	19.8	20.9	0.0	0.0	0.0	0.0	0.0	5.6
(17) = (13) × time cost	Time cost without project	(EUR m)	548.8	17.6	18.7	19.8	20.9	22.2	29.5	39.2	52.2	69.5	87.3
(18) = ((12) − (13)) × other costs	Op. & safety costs of diverted pax	(EUR m)	214.0	0.0	0.0	0.0	0.0	12.5	15.1	18.2	21.9	26.4	28.7
(19)	Appropriated by operator	(EUR m)	0.0	0.0	0.0	0.0	0.0	0.0	0.0	0.0	0.0	0.0	0.0
(20) = − (16) + (17) + (18) − (19)	Net benefit to user	(EUR m)	692.8	0.0	0.0	0.0	0.0	34.7	44.6	57.5	74.2	95.9	110.5
Cost of deterrence											
(21) = 0.5 × (14) × time costs	Time cost with project	(EUR m)	15.7	1.7	1.8	1.9	2.0	0.0	0.0	0.0	0.0	3.6	8.3

Table 4.3   Financial and economic returns of a greenfield airport project *continued*

		Year \ PV	1	2	3	4	5	10	15	20	25	29
(22) = 0.5 × (15) × time costs	Time cost without project	(EUR m)	52.1	1.7	1.8	1.9	2.0	2.1	2.8	3.7	5.0	6.6	8.3
(23) = 0.5 × ((14) − (15)) × other costs	Op. and safety costs of deterred pax	(EUR m)	17.0	0.0	0.0	0.0	0.0	1.2	1.4	1.7	2.1	1.1	0.0
(24)	Appropriated by operator	(EUR m)	0.0	0.0	0.0	0.0	0.0	0.0	0.0	0.0	0.0	0.0	0.0
(25) = − (21) + (22) + (23) − (24)	Net benefit to user	(EUR m)	53.4	0.0	0.0	0.0	0.0	3.3	4.2	5.5	7.0	4.2	0.0
	Cost of congestion												
(26)	With project	(EUR m)	43.0	0.0	0.0	0.0	0.0	0.0	0.0	6.0	7.9	9.6	10.4
(27)	Without project	(EUR m)	0.0	0.0	0.0	0.0	0.0	0.0	0.0	0.0	0.0	0.0	0.0
(28) = − (26) + (27)	Net benefit	(EUR m)	−43.0	0.0	0.0	0.0	0.0	0.0	0.0	−6.0	−7.9	−9.6	−10.4
(29)	Gross producer surplus airport B	(EUR m)	25.0	0.0	0.0	0.0	1.4	1.4	1.7	2.1	2.5	3.0	0.0
(30) = (11) + (20) + (25) + (28) − (29)	Economic flows (ex externalities)	(EUR m)	373.5	−85.0	−85.0	−85.0	−86.4	34.9	46.0	54.5	71.5	88.9	101.5
	ERR without externalities		11.8%										

Externalities

From land transport diversion

(31) = ((13) − (12)) × GHG cost	Greenhouse gases	(EUR m)	19.7	0.0	0.0	0.0	0.8	1.1	1.6	2.2	3.1	3.8
(32) = ((13) − (12)) × noise cost	Noise	(EUR m)	11.9	0.0	0.0	0.0	0.7	0.8	1.0	1.2	1.5	1.6
(33) = ((13) − (12)) × air pollution cost	Air pollution	(EUR m)	17.8	0.0	0.0	0.0	1.0	1.3	1.5	1.8	2.2	2.4
(34) = ((13) − (12)) × safety cost	Safety cost	(EUR m)	11.9	0.0	0.0	0.0	0.7	0.8	1.0	1.2	1.5	1.6
(35) = (31) + (32) + (33) + (34)	Total	(EUR m)	61.3	0.0	0.0	0.0	3.2	4.1	5.1	6.5	8.2	9.3

From generated air transport

(36) = ((15) − (14)) × GHG cost	Greenhouse gases	(EUR m)	58.5	0.0	0.0	0.0	3.1	4.3	6.0	8.4	5.3	0.0
(37) = ((15) − (14)) × noise cost	Noise	(EUR m)	5.7	0.0	0.0	0.0	0.4	0.5	0.6	0.7	0.4	0.0
(38) = ((15) − (14)) × air pollution cost	Air pollution	(EUR m)	3.8	0.0	0.0	0.0	0.3	0.3	0.4	0.5	0.3	0.0
(39) = ((15) − (14)) × safety cost	Safety Cost	(EUR m)	0.6	0.0	0.0	0.0	0.0	0.0	0.1	0.1	0.0	0.0
(40) = (36) + (37) + (38) + (39)	Total	(EUR m)	68.5	0.0	0.0	0.0	3.8	5.1	7.0	9.6	6.0	0.0
(41) = (35) − (40)	Net external effect	(EUR m)	−7.2	0.0	0.0	0.0	−0.5	−1.1	−1.9	−3.1	2.2	9.3
(42) = (30) + (41)	Project net economic flows	(EUR m)	366.2	−85.0	−85.0	−86.4	34.4	45.0	52.7	68.3	91.1	110.8
	Project ERR		**11.6%**									

Should the project be operated by the public sector, total financial returns would include also the taxes collected on inputs and outputs. In town A, the tax rate applying to all revenues and costs, including taxes on energy, sales, etc., as well as social security contributions were fixed at a 15 per cent rate. The resulting tax revenue is included in row 10. Adding tax revenues to operating revenues would still leave the project with a strong negative financial value, as shown in row 11. Moreover, from the point of view of the public sector as a whole, much of the tax revenues would constitute transfers from revenues from airport B. However, airport B falls outside the remit of local authority A, the local authority considering to build airport A, and local authority A disregards financial flows of other local authorities.

Despite the negative financial returns, the project would produce benefits to the local economy, improving the accessibility of local population and local firms, potentially improving the productivity of the local economy. The extent to which this would happen would on average be measured by how much firms and the local population would be willing to pay for the accessibility benefits. This would in turn depend on their willingness to pay for time, which as discussed above was estimated to average €15 per hour. Also as commented above, income per capita in the regional economy was expected to grow by 2 per cent per year and this would be a good approximation to the growth over time in the willingness to pay for time.[7]

Following the surveys conducted, diverted traffic would amount to 600,000 passengers per year and would grow in the future as depicted in row 13. Diverted traffic would travel for 2 hours to town B, incurring time costs, vehicle operating costs and safety costs as depicted in rows 16 to 20. Deterred traffic – potential passengers who do not travel because the generalised cost is too high – with and without the project is shown in rows 14 and 15, respectively. The difference between the two measures

---

[7] As is mentioned in Chapter 2, footnote 2, the assumption about value of time growth must correspond with the assumption made for growth in both labour costs and productivity. For simplicity it is assumed here that labour cost increases are fully compensated by productivity gains.

will constitute traffic generated by the project. So during the first year of operation, in year 5, the airport would generate 132,000 new one-way passenger trips. In year 25, as the airport would get increasingly congested, lack of airport capacity would mean that traffic would start being deterred with the project. In year 29 the project would not generate any traffic, meaning that all traffic flying through the airport would have travelled through airport B, had airport A not been built. The user welfare gain created by generated traffic (or, in other words, net deterred traffic avoided) is estimated in rows 21 to 25, making use of the 'rule of a half', as discussed in Chapter 3, section 1.

Congestion in airport A would become evident from year 15, once traffic exceeds design capacity, as indicated in rows 1 and 2. From there on, service quality at the terminal would diminish, and additional traffic would mean longer queues, waiting times and flight delays. The cost of this congestion is estimated to be around 15 additional minutes of throughput time per passenger and is shown in rows 26 to 28. It is assumed that there is ample capacity in the alternative airport, so no congestion cost is incurred in the 'without project' scenario.[8]

It was seen in the financial appraisal that the local government disregarded the financial effects of the project on region B.

---

[8] As mentioned in footnote 2 in section 1 of Chapter 3, delays caused by congestion may bring about knock on costs to users elsewhere on the air transport network. The US FAA has calculated an average delay propagation multiplier for aircraft of 1.57 for major US airports in 2008 (see FAA 2010). Taking the delay to aircraft as a lower boundary for the delay to passengers, and assuming that it reflects the schedule relationship between the current airport and the rest of the air transport network, the multiplier could be applied to the costs in rows 26 and 28. The effect on this particular project would be small but significant, adding €24.5 million to costs, reducing the NPV by just below 7 per cent to €341.7 million. In this particular example the impact is negative, which is counter-intuitive since it could be expected that a project would alleviate congestion. The reason for the negative result is that in this particular case the alternative airport is assumed to be free of congestion and that it could accommodate all traffic diverted from the project airport. The cost of propagated delay is ignored here because evidence on delays to passengers is not still strong in the literature. However, the cost of delay propagated through the network is bound to become a mainstream item in economic appraisals of aviation projects as more evidence emerges on the magnitude of such costs.

Any private promoter will also follow the same approach. In the economic appraisal, looking at net resource use for the economy at large, all welfare changes need to be included regardless of where they are incurred. Any gross producer surplus (net income plus taxes) that airport B would have generated without the project needs to be subtracted from project benefits. It is assumed that airport B was generating a total of €2 in net producer surplus and tax revenue to the government. This applies only to diverted passengers and the resulting monetary value is included in row 29.[9]

The estimate of economic returns takes into account all monetised effects on the primary market, as picked up in the financial appraisal. Since airport revenues are measured net of taxes, tax revenue has to be included as it constitutes a payment by the user to the promoter that the promoter in turn transfers to the state. Likewise, since both the operating and capital expenditure costs of the airport are measured gross of taxes, tax revenue to the government arising from such items has to be deduced from resource costs and added as a transfer of benefits to the government.[10] Therefore the measure of financial flows to be used in the economic appraisal is as depicted in row 11. Adding to this resource flow both the diversion cost avoided by the project, measured by row 20, the consumer surplus of generated traffic, measured by row 25, and subtracting both the congestion incurred by users of airport A, depicted in row 28, and the reduction in gross producer surplus of airport B, in row 29, widens the measure of resource flow to include welfare changes incurred by the airports and their users, whether monetised or not. It can be seen that the airport generates a net welfare gain of €373.5 million, despite the strong negative financial returns depicted in row 11.

---

9   It is assumed that airport B displays constant returns to scale. Otherwise, a subtraction of passengers may result in higher marginal costs for users of airport B, resulting in a welfare loss, following the analysis in Chapter 2, section 7.3.
10   It is assumed for simplicity that all distortions in the price of inputs concern government taxes.

To complete the estimate of economic return, the analysis needs to include welfare changes to parties not taking part directly in the production or consumption of air transport. The nature of externalities varies depending on traffic type, whether it is generated or diverted. Generated traffic constitutes traffic that would not have travelled without the project and that travels as a consequence of the project. If the passenger paid in full for all external costs of air transport, that is, if all external costs were internalised, there would be no need to make any further adjustments to the economic analysis. Instead, in this case it is assumed that non-internalised environmental costs of air travel include external costs per passenger trip of €20 for greenhouse gas (GHG) emissions, €3 for noise and €2 for air pollution.[11] The marginal cost of GHG emissions is assumed to grow at a cumulative 3 per cent per year, as the cost of additional emissions increases with GHG concentration. In addition, there will be a cost of 30 cents imposed on the rest of society through accident risks.[12] The resulting external costs are measured in rows 36 to 40 in Table 4.3.

Note that it would not be correct to assign such external costs of air transport to diverted traffic, as diverted traffic travels by air in the 'without project' scenario as well, hence such costs cannot be attributed to the project. For diverted traffic, the environmental externality to include would be the external costs inflicted on third parties as a consequence of the diversion. In this case, these would include external costs of the additional road transport necessary to access airport B. Planners estimated that such costs include on a per passenger basis, €1 for greenhouse gases (again growing at 3 per cent per year), €1 for noise, €1.50 for air pollution and €1 as safety costs.[13] Since such costs would be avoided by the project, they represent a project benefit and are included in rows 31 to 35.

---

11 The noise costs are additional to the €20 million already included in the investment cost as the cost of installing double glazing in nearby properties.
12 This risk cost imposed on the rest of society is additional to the €1 cost incurred by the user, included in the estimate of user generalised cost, as shown in Table 4.2 above
13 The €1 of safety cost is additional to the €3 included in the estimate of private generalised cost of transport in Table 4.2 above.

The net external impact is included in row 41 and it is added to the private flows to estimate the overall net economic flows of the project, included in row 42. The economic net present value of the project is €366.2 million and constitutes an economic return on investment of 11.6 per cent. The local government deems that such returns are sufficient to warrant the construction of the airport. However, the strong negative financial return raises the question of how to finance it.

2 Involving the Private Sector (1): No Room for Capital Investment

The results of the project appraisal exercise in Table 4.3 show that the local government of town A faces a potentially viable investment for the economy, producing a sound economic return, but which would yield a strong negative financial return.

The economic analysis shows that most of the value of the project consists of benefits to diverted traffic, totalling €692.8 million, as shown in row 20 of Table 4.3. In addition, there is also the €53.4 million in willingness to pay of generated traffic, shown in row 25. Such consumer surpluses measure the competitive advantage that the airport would have against airport B, in the market for air travellers from region A. One possibility for raising money to help finance the project would be for the airport to appropriate part of that surplus. This could be done by raising aeronautical charges.

The average revenue per passenger (also known as passenger yield) of the airport assumed in the estimation of project viability in Table 4.3 is €8, consisting of €6 in aeronautical yield and €2 in non-aeronautical yield. The estimation of savings in private generalised cost in Table 4.2 shows that the average passenger would value the proximity of airport A by up to €48. That constitutes the upper boundary for an increase in aeronautical charges, since with such an additional charge passengers would be indifferent to whether they travel through airport A or airport B. Planners surveyed other airports elsewhere and found that aeronautical yields of €20 were possible. The local government believed that it would be politically feasible to increase aeronautical yields to €22, which with non-aeronautical yields at €2 would triple passenger yield to €24.

The effect of the pricing policy is depicted in Figure 4.1. The generalised cost to the traveller of travelling through airport A before the increase in airport charges is g_0. The generalised cost of travelling through the alternative airport is g_1, which is €48 higher than g_0. By increasing the aeronautical yield by €16 to €22, the generalised cost of travelling through airport A increases to g_p. The airport sees net operating revenues (i.e. operating revenues minus operating costs) increase by the area $g_p ceg_0$. This is a transfer of surplus from passengers (consumer surplus) to the airport (producer surplus). Because of the increase in prices, traffic falls from q_0 to q_p, which would take the form of lower generated traffic than would be the case without the increase in aeronautical charges.

The effects of increasing aeronautical charges on project returns are shown in Table 4.4. The price increase will result in a small fall in traffic, as depicted in row 2. By year 15, for example, passenger throughput will be about 60,000 lower than with the lower charges. This fall in traffic consists of fewer generated passengers (or more deterred passengers). The higher revenues would reduce substantially the negative return to the operator, increasing the net present value of the investment from a negative €394.6 million to a negative €212.5 million, as shown in row 9 of Table 4.3 and Table 4.4. The higher revenues will also mean higher taxes, as shown in row 10 of both tables. Even so, the project will still have a negative financial worth of €92.8 million to the local government, assuming that the local government operates the airport. That figure would constitute a transfer of money from the taxpayer to the airport user, which is now much smaller than the €304.8 million that would be the case without the increase in charges.

Table 4.4 Financial and economic returns of the greenfield airport project with higher aeronautical charges

			Year\PV	1	2	3	4	5	10	15	20	25	29
	FINANCIAL RETURNS												
(1)	Airport passenger capacity	(thousand)		0	0	0	0	1,200	1,200	1,200	1,200	1,200	1,200
	Passenger throughput												
(2)	With project	(thousand)		0	0	0	0	785	946	1,139	1,373	1,596	1,596
(3)	Without project	(thousand)		0	0	0	0	0	0	0	0	0	0
	Operating cash flows (after tax)												
(4)	With project	(EUR m)	142.1	0.0	0.0	0.0	0.0	6.8	9.2	12.2	15.9	19.8	19.8
(5)	Without project	(EUR m)	0.0	0.0	0.0	0.0	0.0	0.0	0.0	0.0	0.0	0.0	0.0
(6) = (4) − (5)	Net benefit	(EUR m)	142.1	0.0	0.0	0.0	0.0	6.8	9.2	12.2	15.9	19.8	19.8
			0.0										
(7)	Capital investment	(EUR m)	354.6	100.0	100.0	100.0	100.0						
(8)	Subsidy	(EUR m)	0.0	0.0	0.0	0.0	0.0						
(9) = (6) − (7) + (8)	Net cash flow to operator	(EUR m)	−212.5	−100.0	−100.0	−100.0	−100.0	6.8	9.2	12.2	15.9	19.8	19.8
	Operator FRR		N/A										
(10)	Government financial flows	(EUR m)	119.7	15	15	15	15	4	5	6	7	8	8

(11) = (9) + (10)	Operator + government flows	(EUR m)	−92.8	−85	−85	−85	−85	0	0	0	0	0	0	27
	Private and Government FRR		N/A											
	ECONOMIC RETURNS													
	Diverted passengers													
(12)	With project	(thousand)		600	623	646	671	0	0	0	0	0	0	109
(13)	Without project	(thousand)		600	623	646	671	697	839	1,011	1,219	1,469	1,705
	Deterred passengers													
(14)	With project	(thousand)		76	79	82	85	0	0	0	0	0	58
(15)	Without project	(thousand)		76	79	82	85	88	106	128	154	186	216
	Cost of diversion													
(16) = (12) × time cost	Time cost with project	(EUR m)	70.0	17.6	18.7	19.8	20.9	0.0	0.0	0.0	0.0	0.0	5.6
(17) = (13) × time cost	Time cost without project	(EUR m)	548.8	17.6	18.7	19.8	20.9	22.2	29.5	39.2	52.2	69.5	87.3
(18) = ((12) − (13)) × other costs	Op. and safety costs of diverted pax	(EUR m)	214.0	0.0	0.0	0.0	0.0	12.5	15.1	18.2	21.9	26.4	28.7
(19)	Appropriated by operator	(EUR m)	190.2	0.0	0.0	0.0	0.0	11.1	13.4	16.2	19.5	23.5	25.5
(20) = − (16) + (17) + (18) − (19)	Net benefit to user	(EUR m)	502.6	0.0	0.0	0.0	0.0	23.6	31.2	41.3	54.7	72.4	84.9
	Cost of deterrence													
(21) = 0.5 × (14) × time costs	Time cost with project	(EUR m)	9.4	1.1	1.2	1.3	1.3	0.0	0.0	0.0	0.0	1.4	5.5
(22) = 0.5 × (15) × time costs	Time cost without project	(EUR m)	34.7	1.1	1.2	1.3	1.3	1.4	1.9	2.5	3.3	4.4	5.5

Table 4.4 Financial and economic returns of the greenfield airport project with higher aeronautical charges *continued*

		Year \ PV	1	2	3	4	5	10	15	20	25	29
(23) = 0.5 × ((14) − (15)) × other costs	Op. & safety costs of deterred pax	(EUR m) 11.7	0.0	0.0	0.0	0.0	0.8	1.0	1.2	1.4	1.1	0.0
(24)	Appropriated by operator	(EUR m) 20.8	0.0	0.0	0.0	0.0	1.4	1.7	2.0	2.5	2.0	0.0
(25) = − (21) + (22) + (23) − (24)	Net benefit to user	(EUR m) 16.2	0.0	0.0	0.0	0.0	0.8	1.1	1.6	2.2	2.1	0.0
	Cost of congestion											
(26)	With project	(EUR m) 36.1	0.0	0.0	0.0	0.0	0.0	0.0	0.0	7.5	9.6	10.4
(27)	Without project	(EUR m) 0.0	0.0	0.0	0.0	0.0	0.0	0.0	0.0	0.0	0.0	0.0
(28) = − (26) + (27)	Net benefit	(EUR m) −36.1	0.0	0.0	0.0	0.0	0.0	0.0	0.0	−7.5	−9.6	−10.4
(29)	Gross producer surplus airport B	(EUR m) 25.0	0.0	0.0	0.0	1.4	1.4	1.7	2.1	2.5	3.0	0.0
(30) = (11) + (20) + (25) + (28) − (29)	Economic flows (ex externalities)	(EUR m) **364.9**	−85.0	−85.0	−85.0	−86.4	33.9	44.7	58.7	69.5	89.3	101.9
	ERR without externalities	**11.6%**										
	Externalities											
	From land transport diversion											
(31) = ((13) − (12)) × GHG cost	Greenhouse gas	(EUR m) 19.7	0.0	0.0	0.0	0.0	0.8	1.1	1.6	2.2	3.1	3.8

(32) = ((13) − (12)) × noise cost	Noise	(EUR m)	11.9	0.0	0.0	0.0	0.7	0.8	1.0	1.2	1.5	1.6
(33) = ((13) − (12)) × air pollution cost	Air pollution	(EUR m)	17.8	0.0	0.0	0.0	1.0	1.3	1.5	1.8	2.2	2.4
(34) = ((13) − (12)) × safety cost	Safety cost	(EUR m)	11.9	0.0	0.0	0.0	0.7	0.8	1.0	1.2	1.5	1.6
(35) = (31) + (32) + (33) + (34)	Total	(EUR m)	61.3	0.0	0.0	0.0	3.2	4.1	5.1	6.5	8.2	9.3
	From generated air transport											
(36) = ((15) − (14)) × GHG cost	Greenhouse gases	(EUR m)	40.8	0.0	0.0	0.0	2.0	2.9	4.0	5.6	5.3	0.0
(37) = ((15) − (14)) × noise cost	Noise	(EUR m)	3.9	0.0	0.0	0.0	0.3	0.3	0.4	0.5	0.4	0.0
(38) = ((15) − (14)) × air pollution cost	Air pollution	(EUR m)	2.6	0.0	0.0	0.0	0.2	0.2	0.3	0.3	0.3	0.0
(39) = ((15) − (14)) × safety cost	Safety cost	(EUR m)	0.4	0.0	0.0	0.0	0.0	0.0	0.0	0.0	0.0	0.0
(40) = (36) + (37) + (38) + (39)	Total	(EUR m)	47.8	0.0	0.0	0.0	2.5	3.4	4.7	6.4	6.0	0.0
(41) = (35) − (40)	Net external effect	(EUR m)	13.6	0.0	0.0	0.0	0.7	0.6	0.4	0.1	2.2	9.3
(42) = (30) + (41)	Project net economic flows	(EUR m)	378.4	−85.0	−85.0	−86.4	34.6	45.3	59.1	69.6	91.5	111.3
	Project ERR		11.8%									

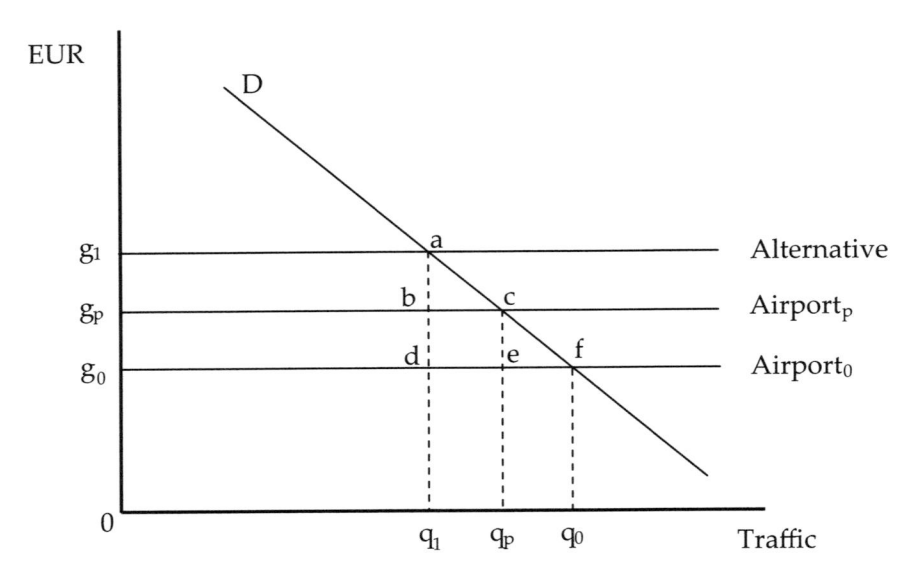

Figure 4.1 Effects of increasing airport aeronautical charges on user and airport surpluses

The loss of welfare to the airport user is registered in row 20 for diverted traffic and row 25 for generated traffic. Note that the economic value of the project before externalities falls slightly from €373.5 million to €364.9 million, as shown in rows 30 of tables 4.3 and 4.4, respectively. The corresponding economic returns fall from 11.8 per cent to 11.6 per cent. There has been a small loss of welfare because some of the traffic generated by the project has now been deterred by the higher airport charges. The loss in economic welfare corresponds to the area cfe in Figure 4.1. However, once externalities are taken into account, the increase in charges actually increases economic value, as depicted in row 42. This is because all of the negative externalities of the project are due to generated traffic, which is reduced by the higher charges. The increase in aeronautical charges can be viewed as a surrogate taxation of externalities, bringing about an improvement in social welfare. Ultimately, however, because the airport project still makes a loss, the broader net effect of the increase in prices is to reduce the transfer of wealth from taxpayers to airport users.

Despite the increase in price, the private sector would not be interested in the project. Whereas aeronautical charges could be increased further, the local government deems it politically unfeasible. The government would still need to operate the project and will expect to incur a financial loss with a present value of €92.8 million after additional tax revenues generated by the project, as shown in row 11.

An attempt to privatise the project would require a transfer of welfare from the public to the private sector. Assuming the private sector demanded a return of at least 8 per cent on the investment, the necessary government subsidy would need to amount to €258.9 million in present value terms, which could take various forms, including a 73 per cent grant on investment costs, or some combination of grants, tax rebates and subsidies to operating costs. None of this would change the economic returns of the project since it would constitute simply a transfer from the public sector to the private sector.

For the project to be worthwhile outsourcing to the private sector, the differences in efficiency of building and operating the project between the private sector and the public sector would have to be substantial. Otherwise the financial return to the private sector will just largely reflect the subsidy. Generally, under such circumstances, private sector involvement in the investment is not justified. Any private sector involvement would instead be through a management contract to operate the airport, minimising or eliminating any upfront capital investment by the private sector and hence minimising transfers from the taxpayer.

This situation is most frequent. Small airports that charge standard levels of aeronautical charges are loss-making, but are economically justified because of non-monetised benefits and are left to the government to develop and operate. There is a rationale for private sector investment in small airports in situations where incomes in the catchment area are very high, increasing user willingness to pay for saving time. In such situations the airport can command high aeronautical charges which, combined with the potential for higher non-aeronautical yields, may constitute investment opportunities with sound financial prospects.

3   Terminal Capacity Expansion

The years following the opening of the airport in town A, the local economy went on to grow faster than had been expected, bringing about faster traffic growth at the airport. It soon became clear that the new airport would become fully utilised earlier than expected. At the originally expected 3.8 per cent growth rate in traffic, the airport terminal was expected to reach design capacity of 1.2 million passengers by year 16, some 12 years after opening. Thereafter it would have continued being able to accommodate additional traffic with some congestion. Deterred traffic would become evident by year 24, and traffic diversion by year 27. Following such projections, it was expected that additional terminal capacity would have to be operational by year 25, some 21 years after the opening of the airport.

Instead of the expected 3.8 per cent growth rate, traffic went on to grow at 5.2 per cent per year. By year 10, passenger throughput was already 1.1 million and the revised demand projection was for traffic growth to average 5 per cent per year over the foreseeable future. Under such projections, the airport would reach design capacity within the next couple of years. Traffic rationing would begin to be evident earlier than expected, traffic deterrence would become evident by year 18, and diversion by year 20.

The airport operator wished to have new capacity in place by year 18 and began preparatory studies for a new terminal, with a view to starting construction no later than year 14. The project would consist of expanding terminal capacity by an additional 1.2 million passengers per year, bringing the total capacity of the airport to 2.4 million passengers. Construction would begin in year 14 and conclude in year 17. Total investment would be €100 million, incurred in equal shares across the four-year construction period. Note that this investment, needed to accommodate an additional 1.2 million passengers, is much smaller than the greenfield investment of €400 million that would also accommodate 1.2 million passengers. Much of the infrastructure capacity developed at greenfield stage, including the runway, apron, tower and access roads, will be sufficient to operate the expanded terminal facility. Since the additional investment is smaller in the expansion stage than in the greenfield stage, but the amount of incremental traffic accommodated is the same in both stages, the returns on investment can expected to be stronger in the expansion stage than in the greenfield stage. Such returns will reflect a mixture of economies of scale and density.

The results of the estimation of financial and economic viability are displayed in Table 4.5. Traffic would reach design capacity of 2.4 million passengers by year 26, some 9 years after opening the expanded facility. Thereafter additional traffic would cause congestion, eventually leading to capacity rationing and deterred traffic by year 32, followed by traffic diversion by year 34, some 17 years after the expanded facility would be opened.

The new project would generate a positive financial return since, unlike at greenfield stage, the present value of the additional operating profits is higher than the investment cost of the project. For the operator the project has a net present value of €37.7 million (row 10) and a financial return of 7.1 per cent. The government will see additional tax revenues with a present value of €53.2 million (row 15). The combined financial value of the project to the operator and the government is 90.9 million (row 16), or a combined financial return of 9.6 per cent.

The economic return is calculated using the same procedure as in the greenfield case, adjusting the 'without project' scenario to include the existing airport capacity in town A and adding a measure of consumer surplus appropriated by the airport from captive traffic (row 34). This is traffic that would have used airport A regardless of whether the project is carried out, that is, total traffic minus deterred and diverted traffic. Such captive traffic was obviously not present at greenfield stage. For ease of comparison with the greenfield case, surplus appropriated by the operator for diverted and deterred traffic is calculated taking as a baseline an airport passenger yield of €8, the same baseline as in the greenfield case. Since the passenger yield is €24, the initial capacity expansion scenario already displays surplus appropriation by the promoter. In the case of captive traffic,

Table 4.5　Financial and economic returns of the terminal capacity expansion project

			Year PV	14	15	16	17	18	25	30	35	42
	FINANCIAL RETURNS											
	Airport passenger capacity											
(1)	With project	(thousand)		1,200	1,200	1,200	1,200	2,400	2,400	2,400	2,400	2,400
(2)	Without project	(thousand)		1,200	1,200	1,200	1,200	1,200	1,200	1,200	1,200	1,200
	Passenger throughput											
(3)	With project	(thousand)		1,374	1,443	1,515	1,591	1,671	2,351	3,000	3,192	3,192
(4)	Without project	(thousand)		1,374	1,443	1,515	1,591	1,596	1,596	1,596	1,596	1,596
	Operating cash flows (after tax)											
(5)	With project	(EUR m)	430.0	16.6	17.8	19.0	20.2	17.4	28.3	39.1	42.7	42.7
(6)	Without project	(EUR m)	303.7	16.6	17.8	19.0	20.2	20.6	20.6	20.6	20.6	20.6
(7) = (5) − (6)	Net benefit	(EUR m)	126.3	0.0	0.0	0.0	0.0	−3.2	7.7	18.5	22.1	22.1
(8)	Capital investment	(EUR m)	88.6	25.0	25.0	25.0	25.0					
(9)	Subsidy	(EUR m)	0.0	0.0	0.0	0.0	0.0					
(10) = (7) − (8) + (9)	Net cash flow to operator	(EUR m)	37.7	−25	−25	−25	−25	−3.2	7.7	18.5	22.1	22.1
(11)	Operator FRR		7.1%									
(12) = (5) − (8) + (9)	Private value of airport with project	(EUR m)	341.4	−8.4	−7.2	−6.0	−4.8	17.4	28.3	39.1	42.7	42.7

			Government financial flows									
(13)		With project	(EUR m)	179.1	10.4	10.7	11.0	11.3	8.5	11.4	14.1	14.9	14.9
(14)		Without project	(EUR m)	125.9	10.4	10.7	11.0	11.3	7.5	7.5	7.5	7.5	7.5
(15) = (13) − (14)		Net revenue	(EUR m)	53.2	0.0	0.0	0.0	0.0	1.0	3.9	6.6	7.3	7.3
(16) = (10) + (15)		Operator + government flows	(EUR m)	90.9	−25	−25	−25	−25	−2.2	11.6	25.1	29.4	29.4
		Private and government FRR		**9.6%**									
		ECONOMIC RETURNS											
		Diverted passengers											
(17)		With project	(thousand)		0.0	0.0	0.0	0.0	0.0	0.0	0.0	206.9	1,590.6
(18)		Without project	(thousand)		0.0	0.0	0.0	0.0	0.0	490.6	1,067.1	1,802.9	3,186.6
		Deterred passengers											
(19)		With project	(thousand)		0.0	0.0	0.0	0.0	0.0	0.0	0.0	430.3	605.4
(20)		Without project	(thousand)		0.0	0.0	0.0	0.0	74.6	264.1	337.1	430.3	605.4
		Cost of diversion											
(21) = (17) × time cost		Time cost with project	(EUR m)	92.9	0.0	0.0	0.0	0.0	0.0	0.0	0.0	9.2	81.4
(22) = (18) × time cost		Time cost without project	(EUR m)	470.3	0.0	0.0	0.0	0.0	0.0	17.9	43.1	80.4	163.2
(23) = ((17) − (18)) × other costs		Op. & safety costs of diverted pax	(EUR m)	157.6	0.0	0.0	0.0	0.0	0.0	8.8	19.2	28.7	28.7
(24)		Appropriated by operator and govt.	(EUR m)	140.1	0.0	0.0	0.0	0.0	0.0	7.8	17.1	25.5	25.5
(25) = − (21) + (22) + (23) − (24)		Net benefit	(EUR m)	237.3	0.0	0.0	0.0	0.0	0.0	18.9	45.2	74.3	84.9

Table 4.5 Financial and economic returns of the terminal capacity expansion project *continued*

		Year \ PV	14	15	16	17	18	25	30	35	42
	Cost of determent										
$(26) = 0.5 \times (19) \times$ time costs	Time cost with project	(EUR m)	34.6	0.0	0.0	0.0	0.0	0.0	0.0	0.0	9.6	15.5
$(27) = 0.5 \times (20) \times$ time costs	Time cost without project	(EUR m)	73.1	0.0	0.0	0.0	0.0	1.2	4.8	6.8	9.6	15.5
$(28) = 0.5 \times ((19) - (20)) \times$ other costs	Op. and safety costs of deterred pax	(EUR m)	18.8	0.0	0.0	0.0	0.0	0.7	2.4	3.0	0.0	0.0
(29)	Appropriated by operator	(EUR m)	33.3	0.0	0.0	0.0	0.0	1.2	4.2	5.4	0.0	0.0
$(30) = -(26) + (27) + (28) - (29)$	Net benefit	(EUR m)	38.5	0.0	0.0	0.0	0.0	0.7	3.0	4.4	0.0	0.0
	Cost of congestion										
(31)	With project	(EUR m)	124.2	5.2	5.5	5.9	6.3	0.0	0.0	15.4	18.1	20.8
(32)	Without project	(EUR m)	111.8	5.2	5.5	5.9	6.3	6.5	7.4	8.2	9.1	10.4
$(33) = -(31) + (32)$	Net benefit	(EUR m)	−12.3	0.0	0.0	0.0	0.0	6.5	7.4	−7.2	−9.1	−10.4
(34)	Surplus appropriated from captive traffic	(EUR m)	0.0	0.0	0.0	0.0	0.0	0.0	0.0	0.0	0.0	0.0
(35)	Gross producer surplus airport B	(EUR m)	17.5	0.0	0.0	0.0	0.0	0.0	1.0	2.1	3.2	3.2
$(36) = (11) + (25) + (30) + (33) - (34) - (35)$	Economic flows (ex externalities)	(EUR m)	**479.9**	−25	−25	−25	−25	4.9	39.9	65.4	91.5	100.8
	ERR without externalities		19.6%									

Externalities

From land transport diversion

(37) = ((18) − (17)) × GHG cost	Greenhouse gases	(EUR m)	470.6	0.0	0.0	0.0	0.0	20.5	51.8	89.8	110.5
(38) = ((18) − (17)) × noise cost	Noise	(EUR m)	26.3	0.0	0.0	0.0	0.0	1.5	3.2	4.8	4.8
(39) = ((18) − (17)) × air pollution cost	Air pollution	(EUR m)	17.5	0.0	0.0	0.0	0.0	1.0	2.1	3.2	3.2
(40) = ((18) − (17)) × safety cost	Safety cost	(EUR m)	2.6	0.0	0.0	0.0	0.0	0.1	0.3	0.5	0.5
(41) = (37) + (38) + (39) + (40)	Total	(EUR m)	517.0	0.0	0.0	0.0	0.0	23.1	57.5	98.3	118.9

From generated air transport

(42) = ((20) − (19)) × GHG cost	Greenhouse gases	(EUR m)	88.9	0.0	0.0	0.0	0.0	2.5	11.1	16.4	0.0	0.0
(43) = ((20) − (19)) × noise cost	Noise	(EUR m)	6.3	0.0	0.0	0.0	0.0	0.2	0.8	1.0	0.0	0.0
(44) = ((20) − (19)) × air pollution cost	Air pollution	(EUR m)	4.2	0.0	0.0	0.0	0.0	0.1	0.5	0.7	0.0	0.0
(45) = ((20) − (19)) × safety cost	Safety cost	(EUR m)	0.6	0.0	0.0	0.0	0.0	0.0	0.1	0.1	0.0	0.0
(46) = (42) + (43) + (44) + (45)	Total	(EUR m)	100.0	0.0	0.0	0.0	0.0	2.9	12.5	18.2	0.0	0.0
(47) = (41) − (46)	Net external effect	(EUR m)	417.0	0.0	0.0	0.0	0.0	−2.9	10.7	39.3	98.3	118.9
(48) = (36) + (47)	Project net economic flows	(EUR m)	896.9	−25	−25	−25	−25	2.0	50.6	104.7	189.8	219.7
	Project ERR		**22.2%**								

surplus appropriation is 0 because the yield of €24 also applies to the 'without project' scenario where there is no capacity expansion.

Costs associated with diverted traffic once more constitute a key determinant of the economic returns of the project (row 22). Note that the analysis assumes that diverted traffic consists of traffic that is forced to make alternative travel arrangements, in this case travelling by car to the nearest alternative airport – the airport in town B, as in the greenfield case – and travelling by air through it. In practice there may be other alternatives. Some travellers may divert to an alternative transport mode to the final destination, although in the case of air transport this would be limited to short-haul traffic.

Likewise, traffic could be diverted to less busy travel times. In particular, once a terminal reaches capacity and airlines are not able to secure slots at preferred times, they may schedule additional capacity at less preferred times. In such cases a demand analysis would have to establish whether this new, less convenient schedule is more attractive to the passenger than travelling through an alternative airport or transport mode. Surveys from passengers can help to shed light on the pattern of passenger behaviour at any airport. However, as far as the calculation of economic benefits is concerned, the additional generalised transport cost the passenger would incur by diverting to the alternative airport constitutes an upper limit to the generalised cost that the passenger would be willing to incur by alternative forms of diversion. That alternative cannot involve a higher generalised cost than travelling through airport B, or else the traveller would travel through airport B.

To the extent that the diverted traveller decides to wait and travel through the airport at a time different from the preferred time, the passenger would generate revenue and operating costs at the project airport. This means that, as far as the financial analysis is concerned, the difference in operating cash flows between the 'with project' and 'without project' scenarios would be smaller than if the passenger diverts to an alternative airport (or transport mode), reducing the financial return of the project. The extent to which the economic (as opposed to financial) return would be affected would depend on the producer surplus assumed in the alternative airport (row 35).[14]

The project generates a strong return of 19.6 per cent before externalities, much higher than in the greenfield case, reflecting the cost efficiencies enjoyed by more efficient use of the airside. Including externalities, the economic return improves even more. This is because the negative external effects created by generated traffic (row 46) are far outweighed by the additional external costs created by land transport used by diverted traffic (row 41). The environmental benefit is larger than in the greenfield case because the real price of carbon is higher, as it is assumed that carbon price would increase by 3 per cent per year in real terms. This is not a necessary result of airport projects. For any project, whether there is a net environmental cost or benefit would depend largely on the means of transport used by diverted traffic. For example, if diversion to the alternative airport is effected by train, which can be far less polluting than road transport, the environmental benefits of avoiding traffic diversion will be smaller and the environmental performance of the project would deteriorate. The result reported here serves only to illustrate that the net environmental impact of an airport is not necessarily negative.

It should be noted that this finding does not run counter to the desirability of internalising the external costs of aviation. Making the users of any transport mode pay for the mode's full external costs yields the most economically efficient outcome. Indeed, if airline users paid for their external costs in full, the economic returns of airport investments would improve, as any non-internalised environmental cost from generated traffic (rows 42 to 44) would disappear.

---

14 The higher the generalised cost of travelling through the alternative airport, the more likely would passengers be to choose to travel through the project airport at less preferred times. Therefore, in cases where the alternative airport constitutes a costly alternative in terms of additional generalised costs, because of being too distant, say, – for example, five hours away – or because significant sea crossings were involved, it would be critical to perform a survey of traveller behaviour since more of the otherwise diverted traffic would travel from the airport at less preferred times.

4 Involving the Private Sector (2): Room for Capital Investment

As is discussed above, unlike the greenfield project the expansion project is profitable, as the expansion exploits economies of scale and density, including the more intensive operation of installed airside infrastructure. In addition, the overall airport now generates a positive financial return and is sellable to private investors as a whole without the need for subsidies or concerns about transfers from the public to the private sector.

The net present value (NPV) of the airport without the project is €303.7 million (row 6). For ease of reference, the analysis ignores expenditure in any refitting that the existing terminal may require. The sale may be made conditional on capacity expansion, even though the private sector would have an incentive to carry out such expansion anyway, because of its expected positive financial return. A private sector operator content with a minimum real return on investment of 5 per cent would be willing to pay the government up to €341.4 million for the airport (row 12), invest the €100 million over the four years required to carry out the terminal expansion project, and generate a 5 per cent return on the total €441.4 million invested. The value of the airport to the private sector would be lower should the risk-adjusted return demanded by the private sector be higher. If, for example, the 7.1 per cent project return represented also the minimum return demanded by the private sector for both the project and the airport as a whole, the value of the airport would be €327.4 million, calculated by discounting the cash flows in row 12 by 7.1 per cent instead of 5 per cent.

Private sector involvement could occur either by an outright sale of the airport, combined with economic regulation if it is judged that there is insufficient competition to check a potential abuse of market power; or by granting a concession for the whole or parts of the airport for a predetermined period.[15]

---

15 It is not the remit of this book to evaluate alternative models of private sector involvement, only to illustrate how investment appraisal plays a role in determining the value of the infrastructure. For a review of models of private sector participation in airports see Winston and de Rus 2008.

Airports 101

For the public sector, the decision to involve the private sector would depend on government budgetary considerations, and on the extent to which any higher cost of capital of the private sector relative to that of the public sector could be expected to be outweighed by a more efficient project implementation by the former. The decision for the private sector to get involved would rest on whether the expected financial returns are sufficient to compensate for the risks included in the competitive, regulatory or contractual frameworks defined by the model of privatisation put forward by the public sector.

In the perhaps simpler case of an outright sale of the airport, government revenues from privatisation would consist of the up to €341.4 million from the sale of the airport (row 12), plus future tax revenues from inputs and output over the life of the project, with a present value of €179.1 million (row 13). This is €53.2 million higher than the €125.9 million the government would raise without the project (row 15). Should the private sector succeed in generating efficiency gains over the life of the project, and should the government decide to pass on some of those efficiency gains to consumers through lower airport charges, tax revenue would decrease through lower input taxes, but increase through any additional traffic generated by the lower charges. Beyond government revenues, however, the lower costs to passengers will bring about a net welfare gain to society, including higher productivity for local businesses, reflected by gains in consumer surplus. Should the airport remain in public sector hands, the project would be worth €90.9 million to the government (row 16), namely the summation of €37.7 million to the (in such a case, public sector) airport operator, plus the €53.2 million net gain in tax revenue.

5 The Incentive to Overinvest

The significant difference between the financial return and the economic return before externalities in the terminal expansion example of Table 4.5 signals a potential for additional revenue generation by the airport. Such potential is measured by non-monetised consumer surplus from diverted traffic (row 25), but not from generated traffic since the level of charges that would appropriate all consumer surplus from diverted traffic would eliminate any generated traffic. In addition, there is potential consumer surplus to be extracted from captive traffic, that is, traffic that would fly from the airport without the project. One way of tapping such revenue would be by increasing charges. But under conditions of economic regulation this avenue would be blocked.[16] Moreover, the regulatory regime will normally include scheduled revisions in charges in order to pass at least part of any efficiency gains on to customers.

One other way of tapping such a consumer surplus would be by overinvesting, or overbuilding. Under a regulated rate of return, profits can be raised by increasing the amount of capital that is remunerated at that rate of return. This incentive to overinvest is called the 'Averch–Johnson effect'[17] and it is illustrated in this section.

Let us assume that the government and the private sector agree to a rate of return on investment of 7.1 per cent – the return arrived at in Table 4.5. The private sector operator could put forward a number of ways to increase investment. It could argue that the traffic forecasts are too conservative, as proved to be the case during the planning of the greenfield project. It could also propose a particularly ostentatious design that would

---

16   The case considered here would not, strictly speaking, be one of economic regulation of monopoly, since there is an alternative airport two hours away. Taking proximity to customer as a service attribute, the market would, rather, be described as one of monopolistic competition. However, economic regulation may still apply on two grounds: first, the infrastructure is operated by the private sector; and second, the market is poorly contestable because of the presence of strong barriers to entry and exit, including sunk costs.

17   The seminal paper, 'Behaviour of the firm under regulatory constraints' by H. Averch and L.L. Johnson (1962), models the incentives faced by firms with an abstract production function including labour and capital. The effect is well known in economics and is widely discussed in utility regulation textbooks. Whereas attempts have been made to de-incentivise such behaviour by firms through the use of price-cap regulation, all price-cap regulation must over the long run pursue attainable rates of return targets in order to incentivise the private sector to invest at all. Therefore, in practice, the Averch–Johnson effect tends to apply to price-cap regulation as well.

Airports

appeal to the public, and in doing so incentivise local politicians to support it. It could also try to convince the authorities to aim for a higher quality of service target than used so far.

Let us assume further that the outcome of such lobbying is to double the size of the airport terminal project, one that would supply additional capacity for 2.4 million passengers per year, instead of the 1.2 million passengers initially envisaged. The new, larger expansion would bring the total capacity of the airport to 3.6 million per year, instead of 2.4 million. The investment cost would double from €100 million to €200 million, and operating costs per passenger in the larger terminal at opening will be 20 per cent higher than in the smaller terminal, due to the lower traffic density.

Table 4.6 shows the effects on project returns of such overinvestment. The financial return would fall to 3.5 per cent from 7.1 per cent. This is because the adverse effects of overdimensioning the terminal, including additional upfront capital investment and additional terminal operating costs, are higher than the favourable effects, which would consist of incremental revenue from extra passengers late into the project life.

The economic return before externalities would fall to 14.4 per cent from 19.6 per cent, as the additional benefits from reducing deterred and diverted traffic far into the future are less than the additional investment and operating costs in the near future. The project generates more value than the smaller project, €562 million versus €479.9 million respectively (row 36 in both tables). However, since the additional value is less than the extra investment required, the rate of return on investment falls. Externalities would also contribute to increasing the value created by the project, as the additional costs, particularly the additional environmental costs brought about by more generated traffic (as generated traffic is now crowded out by diverted traffic later into the future) are smaller than the environmental benefits of reducing the number of passengers diverted by land transport to alternative airports (rows 46 and 41, respectively). The overall return of the project after externalities, at 17 per cent, is still lower than the 22.2 per cent generated with the smaller airport.

Table 4.6 Financial and economic returns of terminal expansion with larger capacity but with no price increase

		Year PV	14	15	16	17	18	25	30	35	42
	FINANCIAL RETURNS										
	Airport passenger capacity										
(1)	With project	(thousand)		1,200	1,200	1,200	3,600	3,600	3,600	3,600	3,600
(2)	Without project	(thousand)		1,200	1,200	1,200	1,200	1,200	1,200	1,200	1,200
	Passenger throughput										
(3)	With project	(thousand)		1,374	1,443	1,515	1,591	1,671	2,351	3,000	3,829	4,788
(4)	Without project	(thousand)		1,374	1,443	1,515	1,591	1,596	1,596	1,596	1,596	1,596
	Operating cash flows (after tax)										
(5)	With project	(EUR m)	433.7	16.6	17.8	19.0	20.2	14.0	24.3	34.6	48.2	64.8
(6)	Without project	(EUR m)	303.7	16.6	17.8	19.0	20.2	20.6	20.6	20.6	20.6	20.6
(7) = (5) − (6)	Net benefit	(EUR m)	130.0	0.0	0.0	0.0	0.0	−6.6	3.7	14.0	27.6	44.2
(8)	Capital investment	(EUR m)	177.3	50.0	50.0	50.0	50.0					
(9)	Subsidy	(EUR m)	0.0	0.0	0.0	0.0	0.0					
(10) = (7) − (8) + (9)	Net cash flow to operator	(EUR m)	−47.3	−50.0	−50.0	−50.0	−50.0	−6.6	3.7	14.0	27.6	44.2
(11)	Operator FRR		3.5%									
(12) = (5) − (8) + (9)	Private value of airport with project	(EUR m)	256.4	−33.4	−32.2	−31.0	−29.8	14.0	24.3	34.6	48.2	64.8

	Government financial flows											
(13)	With project	(EUR m)	212.6	14.2	14.4	14.7	15.1	9.0	12.0	14.8	18.3	22.2
(14)	Without project	(EUR m)	139.2	14.2	14.4	14.7	15.1	7.5	7.5	7.5	7.5	7.5
(15) = (13) − (14)	Net revenue	(EUR m)	73.5	0.0	0.0	0.0	0.0	1.5	4.5	7.2	10.7	14.6
(16) = (10) + (15)	Operator + government flows	(EUR m)	26.2	−50.0	−50.0	−50.0	−50.0	−5.1	8.2	21.3	38.4	58.8
	Private and government FRR		**5.7%**									
	ECONOMIC RETURNS											
	Diverted passengers											
(17)	With project	(thousand)		0.0	0.0	0.0	0.0	0.0	0.0	0.0	0.0	0.0
(18)	Without project	(thousand)		0.0	0.0	0.0	0.0	0.0	490.6	1,067.1	1,802.9	3,186.6
	Deterred passengers											
(19)	With project	(thousand)		0.0	0.0	0.0	0.0	0.0	0.0	0.0	0.0	600.0
(20)	Without project	(thousand)		0.0	0.0	0.0	0.0	74.6	264.1	337.1	430.3	605.4
	Cost of diversion											
(21) = (17) × time cost	Time cost with project	(EUR m)	0.0	0.0	0.0	0.0	0.0	0.0	0.0	0.0	0.0	0.0
(22) = (18) × time cost	Time cost without project	(EUR m)	470.3	0.0	0.0	0.0	0.0	0.0	17.9	43.1	80.4	163.2
(23) = ((17) − (18)) × other costs	Op. and safety costs of diverted pax	(EUR m)	191.9	0.0	0.0	0.0	0.0	0.0	8.8	19.2	32.5	57.4
(24)	Appropriated by operator and govt.	(EUR m)	170.6	0.0	0.0	0.0	0.0	0.0	7.8	17.1	28.8	51.0
(25) = − (21) + (22) + (23) − (24)	Net benefit	(EUR m)	299.7	0.0	0.0	0.0	0.0	18.9	45.2	84.0	169.5

Table 4.6 Financial and economic returns of terminal expansion with larger capacity but with no price increase *continued*

			Year\PV	14	15	16	17	18	25	30	35	42
	Cost of determent											
$(26) = 0.5 \times (19) \times$ time costs	Time cost with project	(EUR m)	6.6	0.0	0.0	0.0	0.0	0.0	0.0	0.0	0.0	15.4
$(27) = 0.5 \times (20) \times$ time costs	Time cost without project	(EUR m)	73.1	0.0	0.0	0.0	0.0	1.2	4.8	6.8	9.6	15.5
$(28) = 0.5 \times ((19) - (20)) \times$ other costs	Op. and safety costs of deterred pax	(EUR m)	29.7	0.0	0.0	0.0	0.0	0.7	2.4	3.0	3.9	0.0
(29)	Appropriated by operator	(EUR m)	52.8	0.0	0.0	0.0	0.0	1.2	4.2	5.4	6.9	0.1
$(30) = -(26) + (27) + (28) - (29)$	Net benefit	(EUR m)	66.5	0.0	0.0	0.0	0.0	0.7	3.0	4.4	6.6	0.1
	Cost of congestion											
(31)	With project	(EUR m)	89.7	5.2	5.5	5.9	6.3	0.0	0.0	0.0	21.8	31.3
(32)	Without project	(EUR m)	111.8	5.2	5.5	5.9	6.3	6.5	7.4	8.2	9.1	10.4
$(33) = -(31) + (32)$	Net benefit	(EUR m)	22.1	0.0	0.0	0.0	0.0	6.5	7.4	8.2	-12.7	-20.8
(34)	Surplus appropriated from captive traffic	(EUR m)	0.0	0.0	0.0	0.0	0.0	0.0	0.0	0.0	0.0	0.0
(35)	Gross producer surplus airport B	(EUR m)	21.3	0.0	0.0	0.0	0.0	0.0	1.0	2.1	3.6	6.4
$(36) = (11) + (25) + (30) + (33) - (34) - (35)$	Economic flows (ex externalities)	(EUR m)	**562.0**	-50.0	-50.0	-50.0	-50.0	2.1	36.6	77.0	112.6	201.2
	ERR without externalities		**14.4%**									

Externalities

From land transport diversion

(37) = ((18) − (17)) × GHG cost	(EUR m)	593.2	0.0	0.0	0.0	0.0	20.5	51.8	101.5	220.6
(38) = ((18) − (17)) × noise cost	(EUR m)	32.0	0.0	0.0	0.0	0.0	1.5	3.2	5.4	9.6
(39) = ((18) − (17)) × air pollution cost	(EUR m)	21.3	0.0	0.0	0.0	0.0	1.0	2.1	3.6	6.4
(40) = ((18) − (17)) × safety cost	(EUR m)	3.2	0.0	0.0	0.0	0.0	0.1	0.3	0.5	1.0
(41) = (37) + (38) + (39) + (40)	(EUR m)	649.7	0.0	0.0	0.0	0.0	23.1	57.5	111.0	237.4

From generated air transport

(42) = ((20) − (19)) × GHG cost	(EUR m)	160.8	0.0	0.0	0.0	2.5	11.1	16.4	24.2	0.4
(43) = ((20) − (19)) × noise cost	(EUR m)	9.9	0.0	0.0	0.0	0.2	0.8	1.0	1.3	0.0
(44) = ((20) − (19)) × air pollution cost	(EUR m)	6.6	0.0	0.0	0.0	0.1	0.5	0.7	0.9	0.0
(45) = ((20) − (19)) × safety cost	(EUR m)	1.0	0.0	0.0	0.0	0.0	0.1	0.1	0.1	0.0
(46) = (42) + (43) + (44) + (45)	(EUR m)	178.3	0.0	0.0	0.0	2.9	12.5	18.2	26.5	0.4
(47) = (41) − (46)	Net external effect	(EUR m)	471.3	0.0	−50.0	−50.0	−2.9	10.7	39.3	84.5	237.0
(48) = (36) + (47)	Project net economic flows	(EUR m)	1,033.3	−50.0	−50.0	−50.0	−0.9	47.3	116.3	197.2	438.3
	Project ERR		17.0%							

Note that tax revenues increase as a result of the overinvestment, as taxes vary directly with investment costs and operating costs, both of which are higher now. This serves to illustrate two issues. First, whereas higher tax revenues are often cited by project promoters as benefits to society, tax revenues and social returns do not necessarily go hand in hand. A project that produces higher tax revenues is not necessarily a project that produces better returns for society if those revenues come from misallocating resources. The second issue is that the government may have a financial incentive to allow overinvestment.

The consequences of allowing overinvestment do not end with the results of Table 4.6. In the example under discussion, government regulation sets a target rate of return for the private sector of 7.1 per cent. For this to occur with the oversized new 2.4 million passenger terminal project, revenue per passenger would have to be increased by 16.3 per cent, relative to that of the 1.2 million passenger terminal project. Since the airport operator has more direct control over aeronautical charges than non-aeronautical revenues, the bulk of the increase would tend to come from increases in charges, with some implications for traffic levels.

Table 4.7 shows the results of the 16.3 per cent charge increase. Whereas the financial return goes up significantly from 3.5 per cent to the targeted 7.1 per cent, the economic return before externalities decreases marginally from 14.4 per cent to 14.3 per cent. The main effects of the price increase are threefold. First, there is a welfare transfer from passengers to both the private operator and, through taxes on revenues, to the government. On the passenger side this is measured by an increase in the appropriation of consumer surplus from diverted traffic from €170.6 million to €212.3 million (row 24), and from captive, or 'existing,' traffic that would have travelled anyway without the project, and which now must pay a higher charge (row 34). From the recipient side, there is an increase in the present value to the private sector from a negative €47.3 million, to a positive €71.5 million (row 10), and of tax revenue to the government from €25.9 million to €163.3 million.

Second, there are changes in resource use, bringing about changes in social welfare, as opposed to just transfers. The higher charges deter some traffic, reducing traffic generated by the project (rows 20 minus 19), although generated traffic remains higher than without overinvestment. On the other hand, less generated traffic reduces congestion costs marginally (row 31). The net effect is a small loss of welfare, consisting of the deadweight loss, identified for the consumer by area cef in Figure 4.1, plus some loss in producer surplus related to deterred traffic. The combination of these losses causes the small fall in economic value generated by the project before externalities from €562 million to €556.8 million (row 36) and in the economic return, relative to the scenario of overinvestment without the price increase.

Finally, there is a change in external effects. The loss of generated traffic improves the environmental and safety performance of the project (row 47), increasing the economic returns marginally to 17.2 per cent up from the 17 per cent achieved in the overinvestment scenario without the price increase.

Comparing Table 4.7 with Table 4.5, the effects of the overinvestment can be summarised as follows. The promoter sees an increase in present value of the investment from €37.7 million to €71.5 million (row 10) while returns per euro invested remain constant. The government sees an increase in tax revenue from €53.2 million to €93 million (row 15). Both the private sector and the government have an incentive to overinvest. Meanwhile, consumers see their surplus affected. Those who would anyway have travelled through the airport without the project have to pay an extra €93.1 million in present value terms to travel (row 34). Travellers who would have diverted without the project see a small improvement in welfare of €20.6 million (€257.9–€237.3 in row 25) as the larger capacity eliminates traffic diversion towards the end of the life of the project, a benefit which is not fully captured by the increase in charges. There is more generated traffic as less traffic is deterred towards the end of the life of the project, but the increase in charges tames the associated welfare gain. In terms of external costs, there is a net improvement as the external benefits from avoiding passenger diversion outweigh the external costs of more generated traffic. All in all, however, whereas society sees extra value created (row 48) it is achieved by devoting disproportionately more resources, resulting in a loss of welfare generation per euro invested, as is evidenced by the decline in economic returns from 22.2 per cent to 17.2 per cent. The less efficient capital allocation should result in lower productivity for the overall economy, subject to the existence of both budget constraints and alternative investment opportunities (which is usually the case).

To the extent that the project has low risks and, in particular, to the extent that the promoter can rely on the willingness of the government to allow the necessary tariff adjustments to maintain the 7.1 per cent return over time, the promoter can further increase returns by leveraging the investment with debt. If the cost of debt financing is less than the 7.1 per cent return on investment, the promoter can debt finance the additional €100 million investment, so that the difference between the cost of debt and the return on investment becomes additional return on equity to the promoter.

The discussion so far helps illustrate the fact that economic regulation, whether through rate of return regulation or through a price cap (with an implicit rate of return target), may not be sufficient to further the interests of society at large. Oversight of capital investment programmes in order to ensure that new capacity is commensurate with reasonable projections of traffic growth may be required. One problem is that such oversight would tend to be carried out through a government agency and, as is shown above, the government may also have an incentive to overinvest because of the positive effects on tax revenues. Therefore, it would be necessary for the agency in charge of approving the investment programme to be kept independent, free from political pressures.

It is important to highlight that the potential for overinvestment ultimately arises from the un-monetised consumer surplus, combined with pricing power by the airport operator. For such pricing power to exist there must be imperfect competition, requiring government regulation, or a strong competitive advantage. Such conditions arise in cases where an airport provides superior accessibility to substantial parts of its catchment area. If competition among airports is close enough to perfect, such room for overinvestment disappears because more efficient capacity planning by the competitor(s) means that the airport that did the overinvesting would experience inferior profitability and would eventually be driven out of business.

Table 4.7 Financial and economic returns of terminal expansion with larger capacity and with price increase

			Year \ PV	14	15	16	17	18	25	30	35	42
	FINANCIAL RETURNS											
	Airport passenger capacity											
(1)	With project	(thousand)		1,200	1,200	1,200	1,200	3,600	3,600	3,600	3,600	3,600
(2)	Without project	(thousand)		1,200	1,200	1,200	1,200	1,200	1,200	1,200	1,200	1,200
	Passenger throughput											
(3)	With project	(thousand)		1,356	1,423	1,494	1,569	1,648	2,318	2,959	3,776	4,788
(4)	Without project	(thousand)		1,374	1,443	1,515	1,591	1,596	1,596	1,596	1,596	1,596
	Operating cash flows (after tax)											
(5)	With project	(EUR m)	552.5	20.9	22.3	23.7	25.2	19.3	31.7	44.0	60.2	80.7
(6)	Without project	(EUR m)	303.7	16.6	17.8	19.0	20.2	20.6	20.6	20.6	20.6	20.6
(7) = (5) − (6)	Net benefit	(EUR m)	248.8	4.3	4.5	4.7	5.0	−1.3	11.1	23.4	39.6	60.1
(8)	Capital investment	(EUR m)	177.3	50.0	50.0	50.0	50.0					
(9)	Subsidy	(EUR m)	0.0	0.0	0.0	0.0	0.0					
(10) = (7) − (8) + (9)	Net cash flow to operator	(EUR m)	**71.5**	−45.7	−45.5	−45.3	−45.0	−1.3	11.1	23.4	39.6	60.1
(11)	Operator FRR		7.1%									
(12) = (5) − (8) + (9)	Private value of airport with project	(EUR m)	375.2	−29.1	−27.7	−26.3	−24.8	19.3	31.7	44.0	60.2	80.7
	Government financial flows											
(13)	With project	(EUR m)	232.2	14.9	15.2	15.5	15.9	9.9	13.2	16.3	20.2	25.0

Table 4.7 Financial and economic returns of terminal expansion with larger capacity and with price increase *continued*

			Year PV	14	15	16	17	18	25	30	35	42
(14)	Without project	(EUR m)	139.2	14.2	14.4	14.7	15.1	7.5	7.5	7.5	7.5	7.5
(15) = (13) − (14)	Net revenue	(EUR m)	93.0	0.7	0.7	0.8	0.8	2.3	5.7	8.8	12.7	17.4
(16) = (10) + (15)	Operator + government flows	(EUR m)	**164.5**	−45.0	−44.8	−44.5	−44.2	1.1	16.8	32.2	52.3	77.6
	Private and government FRR		9.5%									
	ECONOMIC RETURNS											
	Diverted passengers											
(17)	With project	(thousand)		0.0	0.0	0.0	0.0	0.0	0.0	0.0	0.0	0.0
(18)	Without project	(thousand)		0.0	0.0	0.0	0.0	0.0	490.6	1,067.1	1,802.9	3,186.6
	Deterred passengers											
(19)	With project	(thousand)		18.9	19.9	20.9	21.9	23.0	32.4	41.3	52.7	600.0
(20)	Without project	(thousand)		0.0	0.0	0.0	0.0	74.6	264.1	337.1	430.3	605.4
	Cost of diversion											
(21) = (17) × time cost	Time cost with project	(EUR m)	0.0	0.0	0.0	0.0	0.0	0.0	0.0	0.0	0.0	0.0
(22) = (18) × time cost	Time cost without project	(EUR m)	470.3	0.0	0.0	0.0	0.0	0.0	17.9	43.1	80.4	163.2
(23) = ((17) − (18)) × other costs	Op. and safety costs of diverted pax	(EUR m)	191.9	0.0	0.0	0.0	0.0	0.0	8.8	19.2	32.5	57.4
(24)	Appropriated by operator and govt.	(EUR m)	212.3	0.0	0.0	0.0	0.0	0.0	9.8	21.3	35.9	63.5
(25) = − (21) + (22) + (23) − (24)	Net benefit	(EUR m)	257.9	0.0	0.0	0.0	0.0	0.0	17.0	41.0	76.9	157.1

Cost of determent										
(26) = 0.5 × (19) × time costs	Time cost with project	(EUR m)	15.5	0.3	0.3	0.3	0.3	0.4	0.6	0.8	1.2	15.4
(27) = 0.5 × (20) × time costs	Time cost without project	(EUR m)	73.1	0.0	0.0	0.0	0.0	1.2	4.8	6.8	9.6	15.5
(28) = 0.5 × ((19) − (20)) × other costs	Op. and safety costs of deterred pax	(EUR m)	25.5	−0.2	−0.2	−0.2	−0.2	0.5	2.1	2.7	3.4	0.0
(29)	Appropriated by operator	(EUR m)	56.4	−0.4	−0.4	−0.4	−0.4	1.0	4.6	5.9	7.5	0.1
(30) = − (26) + (27) + (28) − (29)	Net benefit	(EUR m)	57.6	−0.1	−0.1	−0.1	−0.1	0.3	1.7	2.7	4.3	0.1
Cost of congestion										
(31)	With project	(EUR m)	81.6	5.1	5.4	5.8	6.2	0.0	0.0	0.0	21.5	31.3
(32)	Without project	(EUR m)	111.8	5.2	5.5	5.9	6.3	6.5	7.4	8.2	9.1	10.4
(33) = − (31) + (32)	Net benefit	(EUR m)	30.2	0.1	0.1	0.1	0.1	6.5	7.4	8.2	−12.4	−20.8
(34)	Surplus appropriated from captive traffic	(EUR m)	93.1	5.4	5.7	5.9	6.2	6.3	6.3	6.3	6.3	6.3
(35)	Gross producer surplus airport B	(EUR m)	21.3	0.0	0.0	0.0	0.0	0.0	1.0	2.1	3.6	6.4
(36) = (11) + (25) + (30) + (33) − (34) − (35)	Economic flows (ex externalities)	(EUR m)	556.8	−50.4	−50.4	−50.5	−50.5	1.5	35.7	75.8	111.2	201.2
	ERR without externalities		**14.3%**									
Externalities										
From land transport diversion										
(37) = ((18) − (17)) × GHG cost	Greenhouse gases	(EUR m)	593.2	0.0	0.0	0.0	0.0	0.0	20.5	51.8	101.5	220.6
(38) = ((18) − (17)) × noise cost	Noise	(EUR m)	32.0	0.0	0.0	0.0	0.0	0.0	1.5	3.2	5.4	9.6

Table 4.7 Financial and economic returns of terminal expansion with larger capacity and with price increase continued

		Year PV	14	15	16	17	18	25	30	35	42
(39) = ((18) − (17)) × air pollution cost	Air pollution	(EUR m)	21.3	0.0	0.0	0.0	0.0	0.0	1.0	2.1	3.6	6.4
(40) = ((18) − (17)) × safety cost	Safety cost	(EUR m)	3.2	0.0	0.0	0.0	0.0	0.0	0.1	0.3	0.5	1.0
(41) = (37) + (38) + (39) + (40)	Total	(EUR m)	649.7	0.0	0.0	0.0	0.0	0.0	23.1	57.5	111.0	237.4
From generated air transport												
(42) = ((20) − (19)) × GHG cost	Greenhouse gases	(EUR m)	139.8	−0.6	−0.6	−0.7	−0.7	1.8	9.7	14.4	21.2	0.4
(43) = ((20) − (19)) × noise cost	Noise	(EUR m)	8.5	−0.1	−0.1	−0.1	−0.1	0.2	0.7	0.9	1.1	0.0
(44) = ((20) − (19)) × air pollution cost	Air pollution	(EUR m)	5.7	0.0	0.0	0.0	0.0	0.1	0.5	0.6	0.8	0.0
(45) = ((20) − (19)) × safety cost	Safety cost	(EUR m)	0.8	0.0	0.0	0.0	0.0	0.0	0.1	0.1	0.1	0.0
(46) = (42) + (43) + (44) + (45)	Total	(EUR m)	154.8	−0.7	−0.7	−0.8	−0.8	2.0	10.9	15.9	23.2	0.4
(47) = (41) − (46)	Net external effect	(EUR m)	494.8	0.7	0.7	0.8	0.8	−2.0	12.2	41.5	87.8	237.0
(48) = (36) + (47)	Project net economic flows	(EUR m)	1,051.7	−49.7	−49.7	−49.7	−49.6	−0.5	47.9	117.3	199.0	438.3
	Project ERR		17.2%									

6 Enlarging a Runway

Runway capacity affects the quality and cost of the air services an airport can offer in two respects. First, runway width and length determine the size of the aircraft the airport can accommodate and whether those aircraft can operate at maximum take-off weight (MTOW). Because larger aircraft have lower operating costs, a larger runway allows airlines to offer services at a lower price per seat or per tonne, or indeed to keep prices unchanged and increase profits, depending on the competitive environment. Also, a longer runway allows airlines to offer longer haul flights, since long-haul routes need heavier take-off weights, if only to carry more fuel. Second, runway capacity determines the maximum number of aircraft movements the airport can accommodate per time, usually measured as movements per hour. This determines both the range of destinations an airport can offer at a given hour and the departure frequency to those destinations, a key determinant of airline schedule quality. It should be noted that beyond the number and size of runways, the runway capacity of an airport is affected by available taxiways, navigational aids, the landscape of surrounding areas (the presence of physical obstacles) and, in airports with more than one runway, by how independently runways can operate from each other.

A runway only rarely constitutes a binding constraint on the passenger capacity of an airport, because limitations on departure frequency can be overcome through increases in the size of aircraft. Runway capacity would constitute a constraint on the passenger throughput capacity of an airport when the runway operates at maximum aircraft movement capacity *and* airlines operate at the highest take-off weight the runway can accommodate. But runway investment projects do not tend to occur in such conditions of absolute necessity. Instead, the decision is based on the willingness to accommodate larger aircraft or to offer greater departure frequency.

This section of the chapter addresses the appraisal of investments to enlarge a runway and section 7, which follows, addresses the case for adding an additional runway. In order to simplify the presentation and help the reader focus attention on air transport issues, the analysis assumes no taxes. The treatment of taxes in economic appraisal is illustrated in the airport terminal case. In addition, in contrast to the greenfield airport and terminal expansion cases, this example will assume that, but for some insulation of nearby houses, which is a typical cost of runway expansion projects, airlines have all their externalities internalised.

The project example consists of the simultaneous widening and lengthening of the existing single runway at an airport. Assume that traffic patterns show that the airport handles an average of 2,000 long-haul passengers a day. The airport does not have a sufficiently large runway to handle International Civil Aviation Organisation (ICAO) Code-D aircraft.[18] As a consequence, the long-haul passengers have to fly to one of three hub airports located one hour's flight away and connect on to intercontinental flights from there.

An airline approaches the airport with a traffic study suggesting that direct flights to the most popular intercontinental destinations could attract 50 per cent of the 2,000 daily long-haul passenger movements currently connecting through one of the nearby hubs, allowing the airport to convert those passenger movements from short-haul to long-haul traffic. For the other 1,000 passengers per day the viable direct flights would not constitute a viable travel alternative. To accommodate the long-haul passenger movements, it will be necessary to enlarge the runway to accommodate Code D operations. A presentation to the airport executives convinces them of the traffic potential and they decide to conduct an appraisal of the investment to check whether it makes financial sense.

For the long-haul passengers originating or ending their trip at the airport, avoiding the connection at any of the three closest hubs would save three hours from the average intercontinental trip. At an average value of time of €15 per hour, saving those three hours would reduce the average behavioural generalised

---

18   Code D aircraft are the smaller of the long- haul, double-aisle aircraft, such as the B-767 or B-787.

cost per one-way trip from €720 to €675. This means that in addition to the 1,000 passengers per day diverted from short-haul connecting flights to long-haul direct flights at a generalised cost elasticity of demand of −1.2, the lower generalised cost could generate a 7.5 per cent increase in traffic, or new trips that would not have taken place without the project.

This generated traffic would account for the main financial gain of the airport. This is because any revenues from the additional charges to the new intercontinental flights would be at the expense of revenues from charges to short-haul connecting flights.[19] On the costs side, the airport would have to invest in lengthening and widening the runway, widening some sections of the existing taxiway (no full parallel taxiway is deemed necessary), and modifying the baggage claim area at the terminal. The capital investment cost at the airport site would be €90 million. In addition, the longer runway would mean that aircraft operations would exceed noise limits for nearby residents, requiring the installation of double glazing in many houses. This would add another €20 million to the cost, bringing the total investment cost of the project to €110 million.

Table 4.8 shows the estimation of project returns, focusing only on the long-haul traffic of the airport that would be affected by the project, initially 1,000 passengers per day, or 365,000 passengers per year. The difference in passenger throughput with and without the project (rows 1 and 2) constitutes traffic generated by the project by reducing the generalised cost of travelling long-haul through the airport. This traffic difference also accounts for the difference between operating cash flows with and without the project (row 5), as the unit costs and unit revenues (or passenger yield) of the airport are the same in both scenarios. The resulting financial return for the airport is strongly negative, with a project NPV of a negative €89.4 million.

---

19 It is assumed that passenger charges are the same for all passengers and that average aircraft landing charge per passenger also works out the same for all aircraft sizes. In reality, aircraft landing charges may be structured in a way that renders the average landing charge per passenger lower for larger aircraft, worsening the financial case for the project.

Table 4.8 Financial and economic returns of a runway enlargement project with no change in aeronautical charges

			Year PV	1	2	3	4	10	15	20	25	28
	FINANCIAL RETURNS											
	Passengers											
(1)	With project	(thousand)		365.0	383.3	402.4	452.7	606.7	774.3	988.2	1,261.2	1,460.0
(2)	Without project	(thousand)		365.0	383.3	402.4	422.5	566.2	722.7	922.3	1,177.2	1,362.7
	Operating cash flow											
(3)	With project	(EUR m)	155.3	5.5	5.7	6.0	6.8	9.1	11.6	14.8	18.9	21.9
(4)	Without project	(EUR m)	146.0	5.5	5.7	6.0	6.3	8.5	10.8	13.8	17.7	20.4
(5) = (3) − (4)	Net benefit	(EUR m)	9.3	0.0	0.0	0.0	0.5	0.6	0.8	1.0	1.3	1.5
(6)	Capital investment	(EUR m)	81.0	20.0	35.0	35.0						
(7)	Intern. of external costs	(EUR m)	17.7	0.0	10.0	10.0						
(8) = (5) − (6) − (7)	Operator financial flows	(EUR m)	−89.4	−20.0	−45.0	−45.0	0.5	0.6	0.8	1.0	1.3	1.5
	Operator FRR		N/A									

ECONOMIC RETURNS

Time benefits											
$(9) = (4) \times$ time costs	To diverted pax	(EUR m)	35.4	0.0	0.0	1.3	2.0	2.9	4.0	5.7	7.0
$(10) = 0.5 \times ((3) - (4)) \times$ time costs	To generated pax	(EUR m)	1.3	0.0	0.0	0.0	0.1	0.1	0.1	0.2	0.2
Lower ticket prices											
(11)	To diverted pax	(EUR m)	0.0	0.0	0.0	0.0	0.0	0.0	0.0	0.0	0.0
(12)	To generated pax	(EUR m)	0.0	0.0	0.0	0.0	0.0	0.0	0.0	0.0	0.0
$(13) = (9) + (10) + (11) + (12)$	Total gain to passengers	(EUR m)	36.7	0.0	0.0	1.4	2.1	3.0	4.2	5.9	7.2
(14)	Profit gain to airline	(EUR m)	465.6	–20.0	–45.0	22.6	30.3	38.7	49.4	63.1	73.0
$(15) = (8) + (13) + (14)$	Economic flows	(EUR m)	**412.8**	–45.0	–45.0	24.5	33.0	42.5	54.6	70.2	81.7
	Project ERR		**23.5%**								

In addition to airport cash flows, an economic appraisal of the investment would also measure non-monetised benefits to passengers, as well as benefits to the airline(s). At the assumed value of time of €15 per hour, project benefits to diverted passengers would have a present value of €35.4 million (row 9). Traffic generated by the project would enjoy a consumer surplus of €1.3 million (row 10).

These gains in consumer surplus to passengers assume that the airline at hand would offer the same ticket price for direct and connecting long-haul flights. However, the airline will experience substantial savings by operating direct long-haul flights, as it will be saving the costs of flying passengers to the connecting hub. Conversations between the airline and the airport reveal that the airline expects such savings to amount to about €50 per passenger. The savings would apply to both diverted and generated traffic and would therefore amount to a very substantial €466.6 million (row 14). This signals that the project generates a lot of value that is not being reflected in the aeronautical revenues of the airport. The project has an economic value of €412.8 million and an economic return of 23.5 per cent. Again, it is assumed that all externalities are internalised. In the event that they were not, the economic return of the project would be higher, as the main project benefits consist of airline operating costs savings by reducing the need for connecting flights to the hub, and airline externalities are directly related to airline output.

The appraisal shows that the airport does not have an incentive to carry out the project, whereas the airline has a strong interest in the project. Clearly, the airline will have an incentive to contribute some of the expected savings of €50 per passenger in order to incentivise the airport to carry out the project. Therefore, the airport suggests to the airline that in order to achieve the 7.1 per cent regulated rate of return on investment, they would need to introduce an increase in landing charges to Code-D aircraft – those benefiting from the project – of €12.8 per passenger, leaving savings in operating costs to airlines at €37.2 per passenger instead of €50.

The implications for project return are shown in Table 4.9. Whereas the increase in charges for Code-D aircraft reduces the benefit to the airline, the project still yields substantial benefits to the airline, amounting to €346.4 million in savings (row 14). This increase in charges is not passed on to passengers and does not change resource use, consisting merely of a transfer from the airline to the airport. Hence whereas the financial value of the project to the airport is now a positive €29.8 million (row 8) and the return on investment is 7.1 per cent, the economic value and economic return remain unchanged after the increase in charges at €412.8 million and 23.5 per cent, respectively.

Strictly speaking, such a scenario would apply to a context of a monopolistic airline market. However, the airline business is competitive and there will be reactions from other airlines. It is not even necessary to assume that other carriers will enter the direct long-haul route, as the market may be too thin to make room for more than one long-haul operator from the airport. But airlines from other competing hubs would lower their prices in order to minimise the loss of business. The airline that approached the airport may also then be forced to lower the price of its air tickets to retain travellers. But the lower time cost to the passenger enabled by the direct service should still make the airline operating the direct service the preferred choice for many passengers. All in all, a plausible final outcome of the project is shown in Table 4.10. The airline passes, say, €20 of its €50 savings in unit costs to passengers, corresponding to a €40 cut on the average return airline ticket price. This generates further traffic, which in turn allows the airport to reduce its required contribution from the airline via higher aircraft charges from €12.8 per passenger to €12, while keeping its 7.1 per cent regulated return on investment. The airline ends up with a net gain of €172.6 million (row 14), down considerably from the €364.4 million that it would have made if the airline industry were not so competitive. Such competition is good for consumers though. Because of the generated traffic from the fare cut, the overall return of the project has increased slightly from 23.5 per cent to €23.8 per cent.

Under an alternative context of political economy, the airline may try to push for the increase in aeronautical charges necessary to finance the project to be spread across all passengers (and airlines) using the airport, irrespective of whether they benefit from the project or not. The airline may claim that the larger runway benefits the local economy by making the region more accessible to the world at large. Lobbyists from the local hotel sector may buy into this argument and support the airline, and politicians may perceive the project as potentially popular. Moreover, the airport operator, rather than limit itself to increasing its regulatory asset base by just €90 million, may take advantage of the political momentum in favour of an investment project and propose an even larger runway expansion project, one suitable for Code-E aircraft, larger than the Code-D needed by the airline, and possibly adding a full-length parallel taxiway. The airline may quietly object to the unnecessary higher cost of upgrading capacity to Code-E rather than the sufficient Code-D, but may decide not to antagonise the airport, to profit from the policy momentum and to accept the higher cost as the price to pay for spreading the charges among all passengers. The final result is that the investment cost will be higher than necessary, the airport operator will make more money by inflating its regulatory asset base (the Averch–Johnson effect, see section 6), the airline will end up paying a slightly higher charge than necessary, although it will not bear the marginal cost of the project, which will be spread across all travellers using the airport. In effect, both the airport and the airline are capturing some of the consumer surplus of all of the passengers using the airport, including those that do not use the long-haul flights prompting the runway extension. As far as society as a whole is concerned, there will be some resource misallocation in the form of a larger runway than would be efficient, as signalled by a fall in the economic return from the project. Such a scenario could only be prevented by effective regulation, including independent oversight of investment plans.

Table 4.9 Returns on a runway enlargement project with change in aeronautical charges and limited airline competition

		Year \ PV		1	2	3	4	10	15	20	25	28
	FINANCIAL RETURNS											
	Passengers											
(1)	With project	(thousand)		365.0	383.3	402.4	452.7	606.7	774.3	988.2	1,261.2	1,460.0
(2)	Without project	(thousand)		365.0	383.3	402.4	422.5	566.2	722.7	922.3	1,177.2	1,362.7
	Operating cash flow											
(3)	With project	(EUR m)	274.5	5.5	5.7	6.0	12.6	16.9	21.5	27.5	35.1	40.6
(4)	Without project	(EUR m)	146.0	5.5	5.7	6.0	6.3	8.5	10.8	13.8	17.7	20.4
(5) = (3) − (4)	Net benefit	(EUR m)	128.5	0.0	0.0	0.0	6.2	8.4	10.7	13.6	17.4	20.1
(6)	Capital investment	(EUR m)	81.0	20.0	35.0	35.0						
(7)	Intern. of external costs	(EUR m)	17.7	0.0	10.0	10.0						
(8) = (5) − (6) − (7)	Operator financial flows	(EUR m)	29.8	−20.0	−45.0	−45.0	6.2	8.4	10.7	13.6	17.4	20.1
	Operator FRR		7.1%									
	ECONOMIC RETURNS											
	Time benefits											
(9) = (4) × time costs	To diverted pax	(EUR m)	35.4	0.0	0.0	0.0	1.3	2.0	2.9	4.0	5.7	7.0
(10) = 0.5 × ((3) − (4)) × time costs	To generated pax	(EUR m)	1.3	0.0	0.0	0.0	0.0	0.1	0.1	0.1	0.2	0.2
	Lower ticket prices											
(11)	To diverted pax	(EUR m)	0.0	0.0	0.0	0.0	0.0	0.0	0.0	0.0	0.0	0.0
(12)	To generated pax	(EUR m)	0.0	0.0	0.0	0.0	0.0	0.0	0.0	0.0	0.0	0.0
(13) = (9) + (10) + (11) + (12)	Total gain to passengers	(EUR m)	36.7	0.0	0.0	0.0	1.4	2.1	3.0	4.2	5.9	7.2

Table 4.9 Returns on a runway enlargement project with change in aeronautical charges and limited airline competition *continued*

			Year \ PV	1	2	3	4	10	15	20	25	28
(14)	Profit gain to airline	(EUR m)	346.4	0.0	0.0	0.0	16.8	22.6	28.8	36.8	46.9	54.3
(15) = (8) + (13) + (14)	Economic flows	(EUR m)	412.8	−20.0	−45.0	−45.0	24.5	33.0	42.5	54.6	70.2	81.7
	Project ERR		23.5%									

Table 4.10 Returns on a runway enlargement project with change in aeronautical charges and competitive airline market

			Year \ PV	1	2	3	4	10	15	20	25	28
	FINANCIAL RETURNS											
	Passengers											
(1)	With project	(thousand)		365.0	383.3	402.4	466.1	624.7	797.2	1,017.5	1,298.6	1,503.3
(2)	Without project	(thousand)		365.0	383.3	402.4	422.5	566.2	722.7	922.3	1,177.2	1,362.7
	Operating cash flow											
(3)	With project	(EUR m)	274.5	5.5	5.7	6.0	12.6	16.9	21.5	27.5	35.1	40.6
(4)	Without project	(EUR m)	146.0	5.5	5.7	6.0	6.3	8.5	10.8	13.8	17.7	20.4
(5) = (3) − (4)	Net benefit	(EUR m)	128.5	0.0	0.0	0.0	6.2	8.4	10.7	13.6	17.4	20.1
(6)	Capital investment	(EUR m)	81.0	20.0	35.0	35.0						
(7)	Intern. of external costs	(EUR m)	17.7	0.0	10.0	10.0						

(8) = (5) (6) − (7)	Operator financial flows	(EUR m)	29.8	−20.0	−45.0	−45.0	6.2	8.4	10.7	13.6	17.4	20.1
	Operator FRR		7.1%									
	ECONOMIC RETURNS											
	Time benefits											
(9) = (4) × time costs	To diverted pax	(EUR m)	35.4	0.0	0.0	0.0	1.3	2.0	2.9	4.0	5.7	7.0
(10) = 0.5 × ((3) − (4)) × time costs	To generated pax	(EUR m)	1.8	0.0	0.0	0.0	0.1	0.1	0.1	0.2	0.3	0.4
	Lower ticket prices											
(11)	To diverted pax	(EUR m)	173.8	0.0	0.0	0.0	8.5	11.3	14.5	18.4	23.5	27.3
(12)	To generated pax	(EUR m)	9.0	0.0	0.0	0.0	0.4	0.6	0.7	1.0	1.2	1.4
(13) = (9) + (10) + (11) + (12)	Total gain to passengers	(EUR m)	220.1	0.0	0.0	0.0	10.3	14.0	18.2	23.6	30.7	36.0
(14)	Profit gain to airline	(EUR m)	172.6	0.0	0.0	0.0	8.4	11.2	14.4	18.3	23.4	27.1
(15) = (8) + (13) + (14)	Economic flows	(EUR m)	**422.4**	−20.0	−45.0	−45.0	24.9	33.7	43.2	55.6	71.5	83.2
	Project ERR		23.8%									

7   Adding a Runway

The previous section of this chapter considers an investment to increase the size and take-off weight of aircraft flying from an airport. This section deals with investments aimed at increasing the number of aircraft movements an airport can handle in a given period of time. The aircraft movement capacity of an airport is generally measured in terms of maximum movements at peak hour, rather than number of movements per day. The types of investments may involve improving the capacity of an existing runway by lengthening a parallel taxiway, adding a second parallel taxiway, adding rapid-exit taxiways or upgrading navigational aids. It can also involve adding a new runway. The analysis below uses as an example the addition of a new runway, but it applies equally to all the investments just mentioned.

Assume there is an airport with a single runway, with a maximum capacity of 50 hourly aircraft movements, 25 take-offs and 25 landings. The runway sees two peak hours a day, one in the morning and one in the evening, Monday to Friday – that is, an average of 260.7 days a year – when it operates close to capacity. Traffic is growing at 4 per cent per year and peak capacity of 50 movements per hour is expected to be reached in three years. The airport managers are considering whether to invest in a second runway. The investment analysis is presented in Table 4.11.

At the moment, airlines operate at the two peak periods with aircraft averaging 100 passengers per flight. This means that in the year, the peak hours see a throughput of about 2.4 million passengers (rows 1 and 15). If a new runway is built, the peak capacity of the airport doubles to 100 movements per hour. In that case, airlines could expand capacity by increasing the number of aircraft movements without needing to change aircraft size. In the long run, however, as larger aircraft are cheaper to operate and as slot availability involves airports at both ends of the route, airlines naturally tend to increase aircraft size as traffic grows. The airport executives calculate that with the new runway, aircraft size will increase on average by some

1 per cent per year, meaning that by year 27, towards the end of the economic life of the project, the average load per aircraft will be 127 passengers (row 2).[20]

The airport managers assume an elasticity of aircraft unit operating cost relative to aircraft size of −0.5, reflecting that larger aircraft are cheaper to operate. This means that by year 27, when the average load per aircraft will be 127 passengers, airline cost per passenger will be 13 per cent lower (row 3). The savings in operating costs to the airlines resulting from using larger aircraft relative to the aircraft used at present would have a present value of €156.1 million (row 9). Should these cost savings be passed on to passengers, there would be some generated traffic. However, since the analysis addresses peak hours under conditions of congestion, it is most probable that the airlines will have pricing power to appropriate such savings, should the airport not appropriate them through higher peak-hour landing charges. The analysis assumes that cost savings are appropriated by the airlines.

In the example at hand, airlines pay for noise externalities via landing charges, but do not pay for greenhouse gas emissions or for air pollution, which remain external costs. Larger aircraft are more fuel efficient on a per passenger basis, so that carrying a given number of passengers on larger rather than smaller aircraft would produce an external environmental benefit by means of reducing emissions. For the average load of 100 passengers per flight the cost of GHG emissions is €20 per passenger. This unit external cost would fall in line with the cost-elasticity of aircraft size of −0.5. In addition, the marginal cost of each tonne of GHG emitted will grow by 3 per cent per year. The combined effect of growing aircraft size through time and increasing marginal cost of GHG emissions produces savings in emissions costs, relative to what would be emitted using current aircraft, with a present value €111.6 million (row 11). In addition, air pollution costs, valued at

---

20 During the first few years after opening a runway the average aircraft size may well drop as airlines schedule more flights in order to secure slots, as long-term investment. The 1 per cent rate of growth of aircraft size would be a long-term assumption.

€2 per passenger, would also fall according to the −0.5 cost-elasticity of aircraft size. Marginal air pollution costs are not assumed to grow through time. This implies that by using larger aircraft there would be savings in emissions costs worth €6.2 million (row 12).

Should the runway not be built, airlines would be further encouraged to increase average aircraft size, as doing so is the only possible way of tapping demand at peak hours. Airport managers assume that, in the absence of the new runway, the airlines would double the rate of increase in average aircraft size from 1 per cent to 2 per cent per year. The average load per aircraft would therefore grow from 100 passengers at present to 161 passengers by year 27 (row 16), rather than to the 127 if the runway was built (row 2). The consequences would be twofold. First, savings in aircraft unit operating costs without the runway would increase to 30 per cent instead of 13 per cent with the new runway. The savings through larger aircraft size would amount to €247.6 million (row 23). Second, there would be further savings in external costs through lower GHG emissions, valued at €174.7 million (row 25); and lower emissions of air pollutants, valued at €9.9 million (row 26).

Those passengers willing to travel during peak hours who could not be accommodated despite the increase in aircraft size would be diverted to alternative departure times. Such traffic diversion can be categorised as frequency delay, in the sense that certain departure times would not be available to (a growing number of) passengers, who will have to travel at less than preferred departure or arrival times.[21] It is assumed that in such cases the frequency delay would be initially one hour as airlines schedule alternative departures at the next best departure/arrival time, namely the hour immediately after or before the peak times. As such demand shoulders become increasingly congested, frequency delay increases.

---

21 Traffic diversion of this type may also be categorised as stochastic delay in the sense that there is no capacity available at the desired flight because of very high load factors. In the case at hand it is just a question of semantics though. The key aspect is that passengers will suffer diversion due to unavailability of flights.

Airport managers estimate that the average delay increases by about 2 per cent per year, half the rate of traffic growth.

Similarly, less departure capacity at peak hours could mean that the number of potential destinations with direct links from/to the airport must be less and, therefore, that a higher proportion of passengers will have to connect through intermediate hub airports. It is estimated that passengers who would be forced to travel through connections at an intermediate hub would incur a loss of two hours relative to a scenario where there are direct links to the airport. The proportion of diverted passengers who travel at alternative times and those diverted to connected routes is dependent on the market conditions of each airport. In this case it is assumed that each constitutes 50 per cent, meaning that initially the average delay for diverted passengers would be 1.5 hours. As the shoulders become more congested, the average delay will grow. The resulting numbers of hours of traffic diversion with and without the projects are included in rows 7 and 21, respectively.

Note that the number of passengers with and without the project is assumed to be the same. Terminal capacity does not constitute a constraint on the project and runways place constraints on aircraft capacity, not necessarily passenger capacity. A runway becomes a constraint on passenger capacity when it is operated at maximum departure frequency *and* at maximum aircraft size. Runway investments are very rarely made when facing such conditions. Rather, they constitute a choice to expand the departure frequency, which has implications for aircraft size, both of which are variables that affect social welfare. Therefore, diverted traffic is assumed to travel from the airport and either travel at alternative times or make connections through hubs.

Differences in traffic diversion and operating costs mean that generalised costs change, hence there is room for generated traffic. Indeed, the analysis could be extended by including generated or deterred traffic. However, unlike the previous examples involving terminals and runways, where the project unambiguously generates traffic, the same cannot be said in the case of an additional runway.[22] This is because when adding a runway, traffic generation occurs both with and without the project. In the 'without project' scenario there will be two factors affecting generated traffic, acting in opposite directions: first, traffic may be diverted by an increasing frequency delay and generalised cost relative to the 'with project' scenario; and second, to the extent that aircraft size increases faster than in the 'with project' scenario, airline ticket prices can be potentially lower, generating traffic. Depending on the strength of these two effects, the 'without project' scenario could result in lower or higher traffic than the 'with project' scenario. Ultimately whether there is net traffic generation with the project relative to the 'without project' scenario rests on the assumptions made about changes in aircraft size in each of the two scenarios. The outcome is very much specific to each project. For simplicity the issue of generated traffic is side-stepped here, which is broadly equivalent to assuming that traffic deterred through greater frequency delay in the 'without project' scenario relative to the 'with project' scenario is offset by traffic generated through higher aircraft size in the 'without project' scenario relative to the 'with project' scenario.

The economic viability of the project would then be determined by a comparison of the investment cost (row 31) with the net savings – aircraft operating costs minus diversion costs – relative to year 1, with the project (row 14) and without the project (row 28). An alternative but equivalent aggregation would be to compare three flows: first, the investment cost of the project (row 31); second, the benefits foregone by the project in terms of lower operating cost and lower external costs that would result from operating larger aircraft, as would be the case without the new runway (rows 23 and 27 minus

---

22 It is worth insisting that we are generally referring to adding new runways to an airport where the existing runway(s) could still accommodate larger aircraft. Where the runway is completely saturated in terms of being unable to accommodate new passengers through larger aircraft, particularly at peak times, a new runway would unambiguously generate traffic.

rows 9 and 13); and third, the project benefits, consisting of the avoidance of costs resulting from passenger diversion to less preferred departure times or routings, thanks to the higher number of aircraft departures allowed by the new runway (row 22 minus row 8). Yet another way to aggregate flows would be to classify benefits into internal (row 29) and external (row 30), and comparing them against investment cost (row 31).

Any of the three ways of aggregation would result in the net project flows as in row 32. The project has an NPV of €172.4 million and an economic rate of return of 11 per cent. Since the assumed opportunity cost of capital is 5 per cent, the project would be viable before any budgeting considerations. Note that the viability of the project rests on internal benefits (row 29), and specifically on benefits to passengers in terms of avoided frequency delay (see rows 8 and 22).[23] However, environmental performance subtracts value from the project (row 30), as the smaller aircraft that would accompany the reduction in frequency delay are more polluting on a per seat basis than the larger aircraft that would operate in the absence of the project.

However, it is worth highlighting the extent to which this result is dependent on the definition of the 'with project' and 'without project' scenarios and, in particular, on the assumed behaviour of airlines in each scenario. The viability of the project rests on two key factors: first, the average size of the aircraft operating in the airport, which determines cost savings through changing the aircraft mix; and second, the diversion cost resulting from fewer frequencies, or frequency delay. Both factors depend on the scheduling practices of airlines. The project analyst must make assumptions about such behaviour and the viability of the project will rely largely on such assumptions.

---

23 Even if operating costs were passed on to passengers, passengers would be willing to trade time savings for higher operating costs, as they value the additional frequency delay (€424.6 million − €1 million = €423.6 million) more than the additional. savings in operating costs (€247.6 million − €156.1 million = €91.5 million).

This is illustrated in Table 4.12, which estimates returns on the same project, assuming that aircraft size in the 'without project' scenario grows somewhat faster, at 2.5 per cent per year instead of 2 per cent per year assumed for Table 4.11. This would mean that by year 27 the average number of passengers per flight on the 'without project' scenario would be 181 passengers (row 16), instead of the 161 assumed previously. The outcome of the 'with project' scenario remains the same (row 14). However, the revised assumption improves the performance in the 'without project' scenario in three ways: first, the faster growth in capacity during peak hours in the 'without project' scenario means that fewer passengers are diverted (row 19), which reduces average frequency delay (row 22); second, the use of larger aircraft increases the savings in operating costs (row 23); and third, the environmental performance improves (row 27). The result is to decrease the NPV of the project decisively, to the point of turning it negative, and to turn the rate of return negative as well.

The key to the outcome of the appraisal rests on the assumption of what airlines would do if there was no additional aircraft movement capacity. Assuming that airlines will not increase aircraft size faster in the 'without project' scenario is not realistic, and would only serve to inflate the estimated returns on investing in a runway.

The revised result may be interpreted as reflecting a situation of strong cost economies, whereby it would make little sense to expand capacity when output is below the minimum efficient scale of capacity already installed. But that does not capture the full nature of the situation. This is because output is not homogeneous, as a second runway improves quality of service by lowering frequency delay. More generally, the viability of an investment on aircraft movement capacity depends on the trade-off between frequency delay and cost savings through aircraft size. This makes both the value of time and aircraft technology central. The higher the value of time, the higher the likelihood that an investment in runway capacity will be economically viable. Therefore, the richer the local economy, the stronger the justification for greater runway capacity for any level of traffic.

This is illustrated in Figure 4.2, which revisits Figure 3.2 and applies it to the case of runways. Schedule C stands for operating costs for a given number of aircraft seats supplied, which varies inversely with aircraft size (AS), and 'FD' stands for frequency delay, the cost caused by not having a departure available at the preferred time. Increasing income increases the value of time, which shifts the frequency delay curve from FD' to FD''. This shifts the equilibrium level of frequency from f' to f''. Frequency level f' is lower than the maximum capacity of a single runway, but frequency level f'' would require a second runway. Therefore, the higher income and accompanying higher value of time makes the case for a second runway even at the expense of higher operating costs resulting from operating smaller aircraft. Note that the total number of passengers does not need to change. The case for a third runway would necessitate much higher increases in income.

Figure 4.2 also illustrates the effect of technology. Whereas in the short- to medium-term, technology – hence the shape of the C curve – can be taken as a given, over the longer run technology that improves aircraft cost efficiency will shift down schedule C. Other things being equal, this would help the case for more runways for any given level of income and traffic. On the other hand, for any level of technology an increase in the cost of fuel or GHG emissions would shift the C schedule upwards, calling for fewer runways for a given level of income and traffic. Advances in aircraft technology tend to be geared towards improvements in fuel efficiency. Therefore, looking to the future, rising incomes and advances in technology can be expected to improve the case for more runways, whereas higher costs of energy and GHG emissions would work against new runways.

Table 4.11 Economic returns of adding a new runway

			Year \ PV	1	2	3	10	20	25	27
With project										
(1)	Demand at peak	(thousand)		2,403	2,503	2,607	3,431	5,078	6,179	6,683
(2)	Pax per aircraft movement	(unit)		100	100	100	107	118	124	127
(3) through (2)	Change in aircraft op. costs	(%)		0%	0%	0%	-4%	-9%	-12%	-13%
(4) through (2)	Capacity at peak	(thousand)		2,607	2,607	5,214	5,590	6,175	6,490	6,621
(5) through (4)	Traffic not diverted	(thousand)		2,403	2,503	2,607	3,431	5,078	6,179	6,621
(6) = (1) − (5)	Traffic diverted	(thousand)		0	0	0	0	0	0	62
(7) = (6) × avge. hours	Hours diverted	(thousand)		0	0	0	0	0	0	147
Internal effects										
(8) = (7) × VoT	Cost of diversion	(EUR m)	1.0	0	0	0	0	0	0	4
(9) through (3)	Aircraft cost savings	(EUR m)	156.1	0	0	0	6	23	38	45
(10) = − (8) + (9)	Internal benefit	(EUR m)	155.1	0	0	0	6	23	38	41
External effects										
(11) through (3)	GHG savings	(EUR m)	111.6	0.0	0.0	0.0	3.3	16.9	31.7	39.7
(12) through (3)	Air pollution savings	(EUR m)	6.2	0.0	0.0	0.0	0.2	0.9	1.5	1.8
(13) = (11) + (12)	External benefit	(EUR m)	117.9	0.0	0.0	0.0	3.6	17.8	33.2	41.5
(14) = (10) + (13)	Net benefit	(EUR m)	272.9	0.0	0.0	0.0	9.8	41.2	71.0	82.4
Without project										
(15)	Demand at peak	(thousand)		2,403	2,503	2,607	3,431	5,078	6,179	6,683
(16)	Pax per aircraft movement	(unit)		100	100	115	140	155	161	

(17) through (16)	Change in aircraft op. costs	(%)		0%	0%	0%	-7%	-20%	-27%	-30%
(18) through (16)	Capacity at peak	(thousand)		2,607	2,607	2,607	2,995	3,651	4,031	4,193
(19) through (18)	Traffic not diverted	(thousand)		2,403	2,503	2,607	2,995	3,651	4,031	4,193
(20) = (15) – (19)	Traffic diverted	(thousand)		0	0	0	436	1,428	2,148	2,489
(21) = (20) × avg. hours	Hours diverted	(thousand)		0	0	0	737	2,940	4,884	5,888
	Internal effects									
(22) = (21) × VoT	Cost of diversion	(EUR m)	424.6	0	0	0	13	64	118	148
(23) through (17)	Aircraft cost savings	(EUR m)	247.6	0.0	0.0	0.0	11.1	36.5	55.0	63.8
(24) = – (22) + (23)	Internal benefit	(EUR m)	–177.1	0	0	0	–2	–28	–63	–84
	External effects									
(25) through (17)	GHG savings	(EUR m)	174.7	0.0	0.0	0.0	6.0	26.4	46.1	56.7
(26) through (17)	Air pollution savings	(EUR m)	9.9	0.0	0.0	0.0	0.4	1.5	2.2	2.6
(27) = (25) + (26)	External benefits	(EUR m)	184.6	0.0	0.0	0.0	6.4	27.9	48.3	59.2
(28) = (24) + (27)	Net benefit	(EUR m)	7.5	0.0	0.0	0.0	4.4	0.1	–14.5	–24.8
	Net project flows									
	Net benefits									
(29) = (10) – (24)	Internal	(EUR m)	332.1	0.0	0.0	0.0	8.3	51.1	100.6	125.0
(30) = (13) – (27)	External	(EUR m)	–66.7	0.0	0.0	0.0	–2.9	–10.0	–15.1	–17.8
(31)	Investment cost	(EUR m)	93.0	50	50					
(32)=(14) – (28) – (31)=(29) + (30) – (31)	Net economic flows	(EUR m)	**172.4**	–50.0	–50.0	0.0	5.4	41.1	85.5	107.2
	Project ERR		**11.0%**							

Table 4.12 Economic returns of adding a new runway with faster growth in aircraft size without the project

		Year \ PV	1	2	3	10	20	25	27
	With project								
(1)	Demand at peak	(thousand)	2,403	2,503	2,607	3,431	5,078	6,179	6,683
(2)	Pax per aircraft movement	(unit)	100	100	100	107	118	124	127
(3) through (2)	Change in aircraft op. costs	(%)	0%	0%	0%	−4%	−9%	−12%	−13%
(4) through (2)	Capacity at peak	(thousand)	2,607	2,607	5,214	5,590	6,175	6,490	6,621
(5) through (4)	Traffic not diverted	(thousand)	2,403	2,503	2,607	3,431	5,078	6,179	6,621
(6) = (1) − (5)	Traffic diverted	(thousand)	0	0	0	0	0	0	62
(7) = (6) × avge. hours	Hours diverted	(thousand)	0	0	0	0	0	0	147
	Internal effects								
(8) = (7) × VoT	Cost of diversion	(EUR m)	1.0	0	0	0	0	0	4
(9) through (3)	Aircraft cost savings	(EUR m)	156.1	0	0	6	23	38	45
(10) = −(8) + (9)	Internal benefit	(EUR m)	155.1	0	0	6	23	38	41
	External effects								
(11) through (3)	GHG savings	(EUR m)	111.6	0.0	0.0	3.3	16.9	31.7	39.7
(12) through (3)	Air pollution savings	(EUR m)	6.2	0.0	0.0	0.2	0.9	1.5	1.8
(13) = (11) + (12)	External benefit	(EUR m)	117.9	0.0	0.0	3.6	17.8	33.2	41.5
(14) = (10) + (13)	Net benefit	(EUR m)	272.9	0.0	0.0	9.8	41.2	71.0	82.4
	Without project								
(15)	Demand at peak	(thousand)	2,403	2,503	2,607	3,431	5,078	6,179	6,683
(16)	Pax per aircraft movement	(unit)	100	100	100	119	152	172	181
(17) through (16)	Change in aircraft op. costs	(%)	0%	0%	0%	−9%	−26%	−36%	−40%

(18) through (16)	Capacity at peak	(thousand)		2,607	2,607	3,099	3,967	4,488	4,716
(19) through (18)	Traffic not diverted	(thousand)		2,403	2,503	2,607	3,099	3,967	4,488	4,716
(20) = (15) − (19)	Traffic diverted	(thousand)		0	0	0	332	1,111	1,690	1,967
(21) = (20) × avge. hours	Hours diverted	(thousand)		0	0	0	560	2,289	3,843	4,653
	Internal effects								
(22) = (21) × VoT	Cost of diversion	(EUR m)	330.5	0	0	0	10	50	93	117
(23) through (17)	Aircraft cost savings	(EUR m)	347.6	0.0	0.0	0.0	14.6	51.7	81.0	95.3
(24) = − (22) + (23)	Internal benefit	(EUR m)	17.1	0	0	0	5	2	−12	−21
	External effects								
(25) through (17)	GHG savings	(EUR m)	247.1	0.0	0.0	0.0	7.9	37.4	67.8	84.7
(26) through (17)	Air pollution savings	(EUR m)	13.9	0.0	0.0	0.0	0.6	2.1	3.2	3.8
(27) = (25) + (26)	External benefits	(EUR m)	261.0	0.0	0.0	0.0	8.4	39.4	71.0	88.5
(28) = (24) + (27)	Net benefit	(EUR m)	278.1	0.0	0.0	0.0	13.0	41.2	59.3	67.1
	Net project flows								
	Net benefits								
(29) = (10) − (24)	Internal	(EUR m)	137.9	0.0	0.0	0.0	1.6	21.7	49.6	62.4
(30) = (13) − (27)	External	(EUR m)	−143.1	0.0	0.0	0.0	−4.9	−21.6	−37.9	−47.1
(31)	Investment cost	(EUR m)	93.0	50	50					
(32)=(14) − (28) − (31)=(29) + (30) − (31)	Net economic flows	(EUR m)	−98.2	−50.0	−50.0	0.0	−3.3	0.1	11.7	15.3
	Project ERR		N/A						

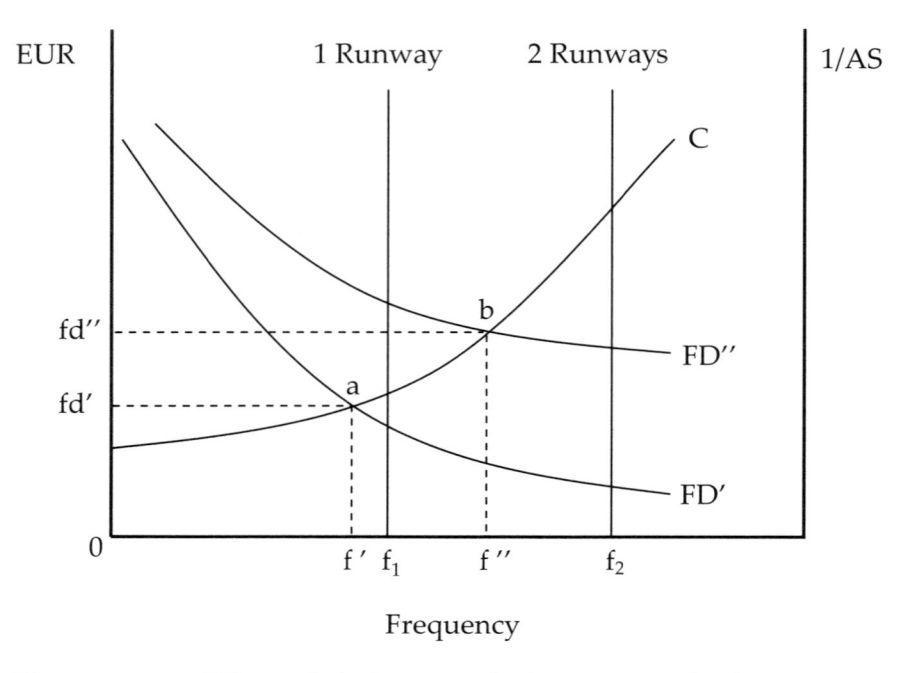

Figure 4.2 Effect of an increase in income on the investment case for a new runway

8 Involving the Private Sector (3): Regulatory versus Competitive Outcome

The above analysis of the investment case for adding a runway does not mention changes in producer surplus, or profitability of the airport operator. Whereas the analysis could have included it, given that the project was assumed not to result in a change in passenger numbers, any change in producer surplus would have reflected largely the structure of airport charges. If the structure of aeronautical charges at the airport were such that the resulting revenue per passenger was constant regardless of the size of aircraft, airport operating revenues would be the same with and without the project.

Any change in airport operating costs would be very case-specific. Airport unit operating costs with the project would increase slightly by operating an extra runway. On the other hand, such costs would be at least in part offset in the 'without project' scenario by the costs of handling larger aircraft, which may require civil works in the terminal, apron and taxiways.

Therefore, given that the project does not necessarily affect passenger throughput, it is quite likely that the financial incentives to the airport for supplying an additional runway would be weak. Moreover, should the airport market be competitive, with no regime of economic regulation by the government, because of the pervasiveness of economies of density in transport a private airport operator may prefer to 'sweat the asset' and squeeze as much traffic as possible through the existing infrastructure. This would call for the airport to show a bias for delaying as long as is possible the building of a new runway.

However, if the airport enjoys some monopoly power and is subject to rate of return regulation, the outcome may differ, for two reasons. First, to the extent that 'sweating the asset' generated super-normal profits, these would be short-lived, as the regulator would subsequently adjust downwards the price cap associated with the regulated rate of return in order to eliminate such super-normal profits. Second, runways are expensive capital assets with a weaker link to traffic than terminals. Since the airport will be remunerated by a predetermined rate of return on regulated assets, the airport would have an incentive to overinvest in airside infrastructure, including runways, for any level of traffic, following the Averch–Johnson effect discussed above in section 6. The implication is that rate of return regulation can create private incentives to show a bias in favour of new runways, even in situations where airport production functions would call for increasing traffic density and operating the existing installed capacity more intensively.

Chapter 5
Air Traffic Management

Introduction

There are two broad types of Air Traffic Management (ATM) infrastructure investments. The first type comprises those aimed at increasing system capacity to handle aircraft movements in a given time period. In terms of investment appraisal, such ATM projects can be approached similarly to an airport investment aimed at increasing the aircraft movement capacity of its runway(s). They would therefore need to incorporate the trade-off between aircraft movement capacity and aircraft size, as airspace capacity constraints can also be partly circumvented by increasing the size of aircraft. The second type of project involves those aimed at improving the efficiency of flight procedures. This project type also has similarities with airport investments, namely those aimed at improving aircraft operations on the ground.[1] This chapter addresses these two ATM project types in turn.

1 Greater Movement Capacity

The treatment of investments aimed at increasing aircraft movement capacity is essentially the same for ATM as for airport runways, facing a trade-off between frequency delay and airport operating costs, as illustrated by Figure 3.2 in Chapter 3. However, there are distinctions to be made depending on the type of airspace sector concerned, whether ground, tower, approach/terminal or en route. Ground control is essentially a component of airport airside operations, aimed at improving capacity before the runway for a given weather condition. Investment in

---

1 In the case of airports this tends not to be a frequent stand-alone investment and is not discussed in Chapter 4, on airports. The treatment for airports would be equivalent to the treatment of flight efficiency in section 3 of this chapter.

such ATM infrastructure is treated essentially as an investment in taxiways. Tower control infrastructure concerns runway operations and its treatment is the same as a runway project aimed at increasing aircraft movement capacity. Approach/ terminal airspace, involving aircraft in and out of airports, to and from their en route sectors, can also be treated as runway investments aimed at increasing aircraft movement capacity.

For en route airspace sectors, an airline is slightly less constrained than when flying in and out of an airport. In the case of insufficient en route airspace capacity, the airline has a choice between diverting to an alternative departure time (similar to the constraints posed by runway capacity), delay on the ground (also similar to the constraints posed by runways), or diversion to alternative en route sectors (different to the constraints posed by runway capacity). If the airline chooses to change route, it will generally involve a longer, and hence more costly, routing than the preferred choice. The costs involved include higher aircraft operating costs for the airline, longer travel time for passengers, and greater greenhouse gas (GHG) emissions externalities if these are not internalised.

Airlines may also react as they respond to constraints posed by runways, by increasing aircraft size. So lack of airspace capacity may also generate benefits – as is the case with runways – by forcing airlines to operate larger aircraft, thereby lowering operating costs per seat and reducing GHG emissions per passenger-kilometre. The impact of air pollutants and noise on en route sectors is debatable. Therefore the treatment of environmental impact in this chapter focuses on GHG emissions.

To summarise, ATM investments aimed at increasing aircraft movement capacity are essentially the same as runway investments with the same aim. The exception would be en route sectors where the airline faces the additional option of altering the route, diverting to alternative sectors.

Another difference between airport and ATM infrastructure investments is that in the latter a greater element of the costs consists of operating costs, instead of capital investment cost, because air traffic controllers can constitute a significant share of the costs of supplying capacity. Since capital investment costs are incurred upfront and operating costs are spread over the operating life of the asset, economic (and hence potential financial) returns on investment will tend to be higher, other things being equal.

Table 5.1 illustrates the estimation of economic returns of a proposed project consisting of upgrading the capacity of an en route sector from 20 to 25 movements per hour, involving investment in IT equipment and employing additional controllers.[2] This capacity limit is reached on working days, three times per day, meaning just over 260 days per year. At an average of 120 passengers per aircraft movement, the capacity of the sector at peak hours would increase from 1.88 million to 2.35 million passengers per year (rows 4 and 19). However, growing traffic means that aircraft size increases with time, increasing the passenger (but not aircraft movement) capacity of the sector through time. As demand is expected to grow faster than aircraft size, planners wish to find out whether it pays to expand movement capacity to 25 movements per hour or whether it is better to signal to airlines that capacity will not be increased in that sector for years, encouraging airlines to increase aircraft size faster.

Analysts estimate that aircraft operating costs are €3,000 per block hour and assume that with the project, aircraft size will increase at 1.5 per cent per year, resulting in an equivalent increase in the average load per flight (row 2). This results in aircraft operating cost savings per seat, governed by an elasticity of unit operating costs relative to aircraft size of −0.5. The analysis assumes that cost savings are not passed on to passengers. The increase in capacity means that traffic diversion will be postponed (row 6), as will the diversion costs incurred by passengers – calculated with a value of time of €15 per hour, growing at 2 per cent per year – (row 8) and the associated additional operating costs to airlines (row 9).

---

[2] For simplicity, the analysis does not include taxes. For an illustration of the treatment of taxes in economic appraisals see the airport terminal cases studied in Chapter 4, sections 4.1 to 4.5.

Table 5.1 Economic returns on an ATM project aimed at increasing aircraft movement capacity

			Year\PV	1	2	3	5	10	15	20	25	27
	With project											
(1)	Demand at peak	(thousand)		1,935	2,016	2,100	2,271	2,763	3,362	4,091	4,977	5,383
(2)	Pax per aircraft movement	(unit)		120	122	124	127	137	148	159	172	177
(3) through (2)	Change in aircraft op. costs	(%)		0%	−1%	−2%	−3%	−7%	−12%	−16%	−21%	−24%
(4) through (2)	Capacity at peak	(thousand)		1,877	1,877	2,346	2,417	2,604	2,805	3,022	3,256	3,354
(5) through (4)	Traffic not diverted	(thousand)		1,877	1,877	2,100	2,271	2,604	2,805	3,022	3,256	3,354
(6) = (1) − (5)	Traffic diverted	(thousand)		58	139	0	0	159	557	1,068	1,721	2,029
(7) = (6) × avg. hours	Hours diverted	(thousand)		18	44	0	0	60	231	488	869	1,065
	Internal effects											
(8) = (7) × VoT	Time cost of diversion	(EUR m)	65.6	0.0	0.0	0.0	0.0	1.1	4.6	10.7	21.0	26.7
(9) through (6)	Op. cost of diversion	(EUR m)	54.4	0.5	1.2	0.0	0.0	1.3	4.6	8.9	14.3	16.9
(10) through (3)	Aircraft cost savings	(EUR m)	70.5	0.0	0.3	0.8	1.6	4.1	6.6	9.3	12.2	13.5
(11) = −(8) − (9) + (10)	Internal benefits	(EUR m)	−49.5	−0.5	−0.8	0.8	1.6	1.7	−2.6	−10.3	−23.1	−30.2
	External effects											
(12)	GHG cost through diversion	(EUR m)	25.1	0.2	0.5	0.0	0.0	0.6	2.1	4.1	6.6	7.8
(13)	GHG savings through aircraft size	(EUR m)	32.4	0.0	0.0	0.4	0.8	1.9	3.0	4.3	5.6	6.2
(14) = −(12) + (13)	External benefit	(EUR m)	7.3	−0.2	−0.5	0.4	0.8	1.3	0.9	0.2	−1.0	−1.6

			-42.3	-0.7	-1.3	1.1	2.4	3.0	-1.7	-10.1	-24.0	-31.8
(15) = (11) + (12)	Net benefit	(EUR m)										
	Without project											
(16)	Demand at peak	(thousand)		1,935	2,016	2,100	2,271	2,763	3,362	4,091	4,977	5,383
(17)	Pax per aircraft movement	(unit)		120	122	125	130	143	158	175	193	201
(18) through (17)	Change in aircraft op. costs	(%)		0%	-1%	-2%	-4%	-10%	-16%	-23%	-30%	-34%
(19) through (17)	Capacity at peak	(thousand)		1,877	1,877	1,877	1,953	2,156	2,381	2,628	2,902	3,019
(20) through (19)	Traffic not diverted	(thousand)		1,877	1,877	1,877	1,953	2,156	2,381	2,628	2,902	3,019
(21) = (16) − (20)	Traffic diverted	(thousand)		58	139	223	318	607	981	1,462	2,075	2,364
(22) = (21) × avg. hours	Hours diverted	(thousand)		18	44	73	108	228	406	668	1,047	1,241
	Internal effects											
(23) = (22) × VoT	Time cost of diversion	(EUR m)	104.7	0.3	0.7	1.1	1.8	4.1	8.0	14.6	25.3	31.2
(24) through (21)	Op. cost of diversion	(EUR m)	95.3	0.5	1.2	1.9	2.7	5.1	8.2	12.2	17.3	19.7
(25) through (18)	Aircraft cost savings	(EUR m)	77.8	0.0	0.5	0.9	1.9	4.4	7.2	10.3	13.7	15.2
(26) = −(23) − (24) + (25)	Internal benefits	(EUR m)	-122.2	-0.8	-1.4	-2.1	-2.5	-4.7	-9.0	-16.5	-28.8	-35.6
	External effects											
(27)	GHG cost through diversion	(EUR m)	44.0	0.2	0.5	0.9	1.2	2.3	3.8	5.6	8.0	9.1
(28)	GHG savings through aircraft size	(EUR m)	36.0	0.0	0.2	0.4	0.9	2.0	3.3	4.8	6.3	7.0
(29) = − (27) + (28)	External benefit	(EUR m)	-8.1	-0.2	-0.3	-0.4	-0.4	-0.3	-0.4	-0.9	-1.6	-2.1
(30) = (26) + (29)	Net benefit	(EUR m)	-130.3	-1.0	-1.7	-2.5	-2.9	-5.0	-9.5	-17.3	-30.5	-37.7

Table 5.1 Economic returns on an ATM project aimed at increasing aircraft movement capacity *continued*

		Year PV	1	2	3	5	10	15	20	25	27
	Net project flows										
(31) = – (8) + (23)	Avoided time costs of diversion	(EUR m)	39.2	0.3	0.7	1.1	1.8	3.0	3.5	3.9	4.3	4.4
(32) = – (9) + (10) + (24) – (25)	Net aircraft cost savings	(EUR m)	33.5	0.0	–0.1	1.7	2.4	3.4	2.9	2.3	1.5	1.1
(33) = (14) – (29)	GHG savings	(EUR m)	15.3	0.0	–0.2	0.8	1.1	1.6	1.4	1.1	0.7	0.5
(34) = (31) + (32) + (33) = (15) – (30)	Differential net benefits	(EUR m)	88.0	0.3	0.4	3.6	5.3	8.0	7.8	7.3	6.4	6.0
(35)	Increase ATC op. cost	(EUR m)	3.8	0.0	0.0	0.3	0.3	0.3	0.3	0.3	0.3	0.3
(36)	Investment cost	(EUR m)	13.9	7.5	7.5							
(37) = (34) – (35) – (36)	Net economic flows	(EUR m)	70.3	–7.2	–7.1	3.3	5.0	7.7	7.5	7.0	6.1	5.7
	Project ERR		32%									

The use of larger aircraft will reduce GHG emissions per passenger. At 120 passengers per flight, GHG costs are estimated at €1,386 per block hour.[3] Emissions are assumed to vary proportionally with aircraft operating costs. Aircraft emissions costs savings per passenger through the use of larger aircraft relative to the situation in year 1 are included in row 13. The project also postpones GHG emissions costs associated with diverted traffic (row 12).

Without the project, analysts assume that airlines will increase aircraft size faster, at 2 per cent per year, rather than 1.5 per cent with the project. Whereas this mitigates the capacity shortage, more traffic is diverted without the project (row 21) than with the project (row 6). The larger aircraft also produce cost efficiency gains relative to the project scenario (rows 25 and 10). However, since more traffic is now diverted, higher operating costs (row 24), time costs (row 23) and GHG emissions costs (row 27) are borne without the project.

The net benefits of the project will depend on the extent to which the lower diversion costs and lower aircraft operating and emissions costs through more direct routing with the project are offset by the use of larger aircraft without the project. On balance the project improves the performance of air transport on all relevant counts. It produces net savings in diversion costs to passengers of €39.2 million (row 31). The shorter routing with the project outweighs the operating cost penalty of lower aircraft size, resulting in net aircraft operating cost savings of €33.5 million (row 32). Lower aircraft operating costs also translates in emissions savings worth €15.3 million (row 33).

The project costs to the air navigation service provider (ANSP) consist of capital investment in equipment with a present value of €13.9 million (row 36), and an increase in operating costs by

---

3 The analysis assumes a flat real cost of GHG emissions of €35 per tonne of emissions throughout the project life. This is an alternative approach to the airport runway case, where a cost of €20 per tonne in year 1 is assumed, growing at 3 per cent per year throughout the life of the project. A discussion of the merits of alternative scenarios regarding future emissions costs is beyond the scope of this book. The important point made here is that the project appraisal must include external costs.

€300,000 per year, mostly by requiring more controller hours. The net flows of the project are included in row 37. The project does produce a net benefit worth €70.3 million, constituting a very strong return on capital investment of 32 per cent.

Note, however, that towards the end of the project life benefits decrease with time, as the cumulative effect of increasing aircraft size over time begins to yield significant differences in operating costs. Indeed, a lot depends on the assumed scenarios regarding aircraft size increases with and without the project. Should the analyst assume that aircraft size with the project will increase at 1 per cent per year, instead of the assumed 1.5 per cent, the returns from the project would decrease from 32 per cent to 6 per cent. If, in addition, the analyst assumes that aircraft size without the project will increase at 2.5 per cent instead of 1 per cent, then the project will generate negative returns. As in the case of runways, the benefits of investments in increasing ATM capacity rely substantially on assumptions about the ability of airlines to accommodate traffic growth through larger aircraft. The same cannot be said about ATM investments aimed at enabling more efficient operation of aircraft, as can be seen in the project example in section 4 below.

Note that so far, and as was the case with the runway project, the analysis does not include producer surplus. In effect, the analysis assumes that ANSP revenues are exactly the same with and without the project and therefore cancel out. This involves two assumptions. First, it is assumed that with and without the project the same number of passengers is served. This is not necessarily a controversial assumption, as the direction of any generated traffic is not obvious. It would result from the balance between the cost of diversion to alternative sectors and the benefits of lower operating costs through larger aircraft (should such costs be eventually passed on to passengers). On the other hand, and regarding the financial analysis, it may well be that traffic is diverted to a sector managed by a separate ANSP, in which case there would be a loss of revenue to the promoter. The next section of this chapter, dealing with private sector involvement, addresses this issue.

Air Traffic Management 149

The second assumption is that the air navigation charge per passenger is the same with and without the project. This may well be the case – as when air navigation is paid as a levy on ticket price – but it is generally not so in practice. The implications are discussed in the next section.

2 Involving the Private Sector (4): Pricing Policy

ANSPs are mostly operated by the public sector, even if they are 'corporatised', and when they are privatised they are operated as regulated monopolies. Beyond any rate of return regulation, which, as seen in Chapter 4, section 5, incentivises overinvestment, the pricing structure may itself affect incentives to invest for a private sector ANSP.

ANSPs usually follow International Civil Aviation Organisation (ICAO) guidelines regarding air navigation charges, structuring them according to both route length and aircraft size. The precise implementation of such guidelines varies across ANSPs, though. Some apply formulas whereas others determine price lists organised by ranges of flight distance and aircraft weight. Other ANSPs are remunerated through a levy set as a percentage of air ticket price or per flight. Where formulas are used, an example may be the following, used by Eurocontrol:

$$Charge = Unit\ Rate \times \frac{Distance}{100} \times \left(\frac{MTOW}{50}\right)^n$$

… where the unit rate is a constant, measured in the applicable currency; route length is measured as the great-circle distance in kilometres between the two extremes of the airspace section where the ANSP controls the flight; the weight is measured by the aircraft's maximum take-off weight (MTOW); and n manages the proportionality between aircraft weight and the charge. The air navigation charge increases with distance and with aircraft weight, meeting ICAO recommendations.

Assuming a unit rate of €50, an MTOW of 75 tonnes for a Code-C aircraft, and n=1 (not necessarily the factor used by Eurocontrol), the charge applicable to the average flight in the ATM project example in section 1 above would be:

$$Charge = 50 \times \frac{1,000}{100} \times \left(\frac{75}{50}\right)^1 = EUR\,750.00$$

The €750 charge would be paid by the airline irrespective of the number of passengers on board. However, setting an average charge per passenger is useful for the discussion at hand. Following the assumption in the project example that the flight carries on average 120 passengers in year 1, the ATM charge per passenger would be €6.25.

The price formula is such that an increase in the aircraft size, carrying on average more passengers, would result in a higher charge overall for the flight, but a lower charge per passenger. For example, if the aircraft can carry an extra 25 passengers at the same load factor and has an MTOW higher by 5 tonnes, the resulting total charge would be €800 and the charge per passenger €5.52.

Table 5.2 uses the example in Table 5.1 to simulate the effect on ANSP revenues of applying a formula of this type. To simplify, it is assumed that aircraft technology and the n exponential factor applicable are such that as aircraft size increases, the resulting air navigation charge per passenger decreases by half the percentage saving in aircraft unit operating costs.[4] The table measures the increases in revenues relative to revenues in year 1 with and without the project. The difference between the increase in revenues with and without the project would then measure net revenue increase, which is determined only by the changes in the charge applicable to traffic, since the total amount of traffic does not change.

---

4 Since it is assumed in turn that the elasticity of aircraft unit operating costs with respect to aircraft size is −0.5, the parameters of the price formula are such that, say, a 10 per cent increase in aircraft size, resulting in a 5 per cent fall in aircraft unit operating cost, would cause a 2.5 per cent fall in the applicable air navigation charge per passenger.

Table 5.2 Financial returns for an ANSP of investing in greater aircraft movement capacity

			Year\PV	1	2	3	5	10	15	20	25	27
	DIVERTED TRAFFIC STAYS WITH THE SAME ANSP											
	With Project											
(1)	Incr. revs. from traffic not diverted	(EUR m)	59.7	0.0	0.0	1.4	2.4	4.4	5.5	6.6	7.7	8.1
(2)	Incr. revs. from diverted traffic	(EUR m)	35.5	0.0	0.5	−0.4	−0.4	0.6	3.1	6.3	10.4	12.3
(3) = (1) + (2)	Total incremental revenues	(EUR m)	95.2	0.0	0.5	1.0	2.1	5.0	8.6	12.9	18.1	20.5
	Without project											
(4)	Incr. revs. from traffic not diverted	(EUR m)	29.5	0.0	0.0	0.0	0.5	1.7	2.9	4.2	5.4	5.9
(5)	Incr. revs. from diverted traffic	(EUR m)	66.2	0.0	0.5	1.0	1.6	3.4	5.8	8.8	12.6	14.4
(6) = (4) + (5)	Total incremental revenues	(EUR m)	95.8	0.0	0.5	1.0	2.1	5.1	8.7	12.9	18.0	20.3
	Net project flows											
(7) = (3) − (6)	Net revenue gain	(EUR m)	−0.6	0.0	0.0	0.0	0.0	−0.1	−0.1	0.0	0.0	0.1
(8)	Increase ATC op. cost	(EUR m)	3.8	0.0	0.0	0.3	0.3	0.3	0.3	0.3	0.3	0.3
(9)	Investment cost	(EUR m)	13.9	7.5	7.5							
(10) = (7) − (8) − (9)	Financial flows	(EUR m)	−18.3	−7.5	−7.5	−0.3	−0.3	−0.4	−0.4	−0.3	−0.3	−0.2
	FRR		N/A									
	DIVERTED TRAFFIC FLIES THROUGH AN ALTERNATIVE ANSP											
(11) = (1) − (4)	Rev. gain from non-diverted traffic	(EUR m)	30.1	0.0	0.0	1.4	2.0	2.7	2.6	2.4	2.3	2.2
(12) = (11) − (8) − (9)	Financial flows	(EUR m)	12.4	−7.5	−7.5	1.1	1.7	2.4	2.3	2.1	2.0	1.9
	FRR		11.7%									

The resulting financial return of the project, measured against what would have happened in the 'without project' scenario, is negative. The operator does not have an incentive to invest, even though – it is important to recall – the project has a strong economic return of 32 per cent (see Table 5.1). Note that any gain or loss in revenue would come from a change in price and would constitute a transfer between the ANSP and the airline or the passenger and, therefore, does not represent a net change in welfare to be added to the calculation of economic return.[5]

The lower part of Table 5.2 calculates the same scenario, but assuming that diverted traffic flies instead through a sector managed by an alternative ANSP. Now, obviously, not carrying out the project would result in 'lost customers' and therefore the financial return of the project increases and in the current case becomes a positive 11.7 per cent. As far as the economic appraisal is concerned, the producer surplus of the alternative ANSP is treated equally to that of the promoter at hand, and therefore can again be ignored.

Therefore, other than when traffic would divert to an alternative ANSP, the ANSP does not have an incentive to invest, even when the project produces strong economic returns. This outcome is the product largely of the pricing policy. Economic efficiency would call for prices to be set at marginal cost which, on a long-term perspective, can be taken to mean average cost. The cost that a controlled flight causes to the ANSP depends on the amount of time it needs to be controlled and, therefore, on distance and speed. Jet passenger aircraft of different sizes travel at similar speeds and require the same amount of workload and resources from the ANSP. Therefore economic efficiency would call for jet aircraft to be charged the same amount regardless of their size. Instead, by applying common pricing policy whereby charges increase with aircraft weight, passengers in larger aircraft may end up paying less for air navigation services than passengers in smaller aircraft, although they are still cross-subsidising them.

---

5 Only a price change resulting in generated traffic would generate a change in welfare to be added to the economic returns. The resulting relative price changes to airline tickets, however, tend to be small, so that any generated or deterred traffic can also be expected to be small.

The distortion is more acute with smaller propeller aircraft. They tend to be slower and hence require more controller workload per distance travelled; they can also occupy a given section of airspace for longer than a jet aircraft. Economic efficiency would require slower propeller aircraft to be charged more than jet aircraft. Instead they are usually charged less.

Putting aside the issue of propellers and remaining with jets, if prices were set efficiently, by speed and distance rather than weight and distance, the 'without project' scenario in the example in Table 5.2 would have seen a substantial fall in revenues, because the ANSP would not be able to price-discriminate against passengers on larger aircraft. The result would be to improve the financial performance of the project, encouraging the ANSP to invest in more capacity.

By enabling ANSPs to price-discriminate against larger aircraft, ICAO may be adding an element of solidarity into paying for aircraft services. But it is also reducing the incentive to expand capacity by allowing ANSPs to profit from passengers switching to larger aircraft. It may be disincentivising investment even in cases where, as the example above suggests, the investment would produce strong economic returns.

In practice, however, ANSPs operated by the private sector would have their price caps tied to rate of return regulation, which in turn is tied to the asset base of the operator. As is seen in Chapter 4, section 5, rate of return regulation incentivises investment, and indeed overinvestment, in capacity. In the case of ATM, increases in flight movement capacity generally go hand in hand with a larger asset base. Therefore, rate of return regulation would incentivise investment that the current pricing policy disincentivises. On the other hand, not all asset base increases involve increases in capacity. They may also improve safety, the quality of communications, reduce controller or pilot workload, and so on.

To conclude, a private ANSP operating under rate of return regulation and with economically efficient prices would have strong incentives to invest in added capacity. The current pricing policy of discriminating against passengers in larger aircraft somewhat diminishes that incentive, even in cases where capacity expansion is justified. Rate of return regulation would tend to ensure that investment is always forthcoming. On the other hand, the incentive to invest would not necessarily be in increasing capacity but rather on upgrading technology.

3   Flight Efficiency

This section addresses projects that are aimed not at increasing capacity but at making existing flights more efficient, generally by making more direct flights (horizontal efficiency), and more smooth, uninterrupted climbs and descends (vertical efficiency). Efficiency and capacity are not independent from each other. Routes that are more direct can also increase capacity by minimising the use of airspace and controller input. Also, during busy periods, attempting to improve efficiency of individual flights can penalise system capacity by imposing constraints on other flights.

For clarity, this section of the chapter addresses a project aimed exclusively at improving flight efficiency, with no implications for capacity. An ATM project aimed at improving flight efficiency but with a knock-on impact on capacity would also need to incorporate the appraisal approach discussed in section 1 above. It should be noted that the method presented here could also be applied to projects aimed at improving aircraft operations at airports, such as new taxiways that diminish taxying distance and time.

The project consists of the installation of ground navigational aids and IT equipment to enable an airport and the associated approach ANSP to offer airlines continuous climb departures (CCD) and continuous descend approaches (CDA) – the latter are also known as optimised profile descent (OPD) in the US. Both procedures improve the vertical efficiency of aircraft operations, minimising the need for level segments at altitudes lower than cruising, where flying is more expensive in terms of fuel burn. For the airport at hand, the CCD is estimated to reduce fuel burn by 10 per cent, and the CDA by 40 per cent in the climb and descent segments, respectively. The optimised procedures are expected to apply to 15 per cent of flights, occurring at off-peak hours, or about 10 departures and 10 arrivals a day.

Fuel is assumed to cost €600 per tonne, including €105 as the cost of GHG emissions, which are fully internalised through airline ticket prices.[6] Before the project the average departure is assumed to consume 2 tonnes of fuel and the average arrival 250 kg. The benefits in terms of lower operating costs would accrue to the airlines and, if markets are sufficiently competitive, would eventually be passed on to the passengers. The greater number of efficient procedures would also reduce noise impact and improve air quality in the vicinity of the airport, externalities that are not internalised, unlike GHG emissions. The impact of noise is currently estimated to average €100 per aircraft movement, and the improved procedures to reduce the incidence by 20 per cent. The cost in terms of local air quality is estimated at €125 per aircraft movement and the project would reduce the incidence by 15 per cent. The benefits from reducing these externalities accrue to residents in the airport's vicinity.

In projects aimed at increasing ATM capacity discussed in section 1 above, the analyst has to make assumptions about airline behaviour in the 'with project' and, perhaps more critically, in the 'without project' scenarios. If a project is not carried out airlines may chose alternative routings or larger aircraft sizes and passengers may decide on alternative routings or departure times. These assumptions that the analyst must make about airline behaviour are not self-evident, yet they can be decisive for the estimated returns of the project. In the case of projects aimed only at improving flight efficiency there is no need to make assumptions about passenger or airline behaviour in a 'without project' scenario. The 'without project' scenario is simply the current situation.[7]

---

6 GHG emissions are assumed to be 3 tonnes of GHG per tonne of fuel and are priced at €35 per tonne in real terms throughout the life of the project. Neither of the two assumptions – price and internalisation – hold at the time of writing. They are used in order to illustrate the treatment of alternative regulatory circumstances in project appraisal. See previous project examples for the treatment of alternative market and regulatory contexts.

7 If the airline market were competitive and the cost savings were passed on to passengers, the project might generate traffic. In that case, traffic

Table 5.3   Economic returns on an ATM investment project aimed at improving flight efficiency

		Year \ PV	1	2	3	4	5	10	15	20	21
(1)	Approaches	(unit)	3,650	3,796	3,948	4,106	4,270	5,195	6,321	7,690	7,998
(2)	Departures	(unit)	3,650	3,796	3,948	4,106	4,270	5,195	6,321	7,690	7,998
(3) through (1)	Fuel saved in approaches	(tonnes)	0	380	395	411	427	520	632	769	800
(4) through (2)	Fuel saved in departures	(tonnes)	0	759	790	821	854	1,039	1,264	1,538	1,600
	Internal benefits										
(5) through (3) and (4)	Value of fuel saved	(EUR k)	11,335	0	683	711	739	769	935	1,138	1,384	1,440
	External benefits										
(6) through (1) and (2)	Air pollution benefits	(EUR k)	2,362	0	142	148	154	160	195	237	288	300
(7) through (1) and (2)	Noise benefits	(EUR k)	2,519	0	152	158	164	171	208	253	308	320
(8) = (5) + (6) + (7)	Total benefits	(EUR k)	16,216	0	977	1,017	1,057	1,100	1,338	1,628	1,980	2,059
(9)	Investment costs	(EUR k)	5,714	6,000								
(10) = (8) – (9)	Net economic flows	(EUR k)	**10,501**	–6,000	977	1,017	1,057	1,100	1,338	1,628	1,980	2,059
	Project ERR		**19%**									

Table 5.3 presents the key input variables and the result. The benefits consist of fuel saved by the airlines (row 5), and lower air pollution and noise to residents in the vicinity of the airport (rows 6 and 7, respectively). The costs consist of the capital investment (row 9). No significant operating cost differences are expected and the operating life of the equipment installed is expected to be 20 years. The investment is clearly viable, with a strong economic return of 19 per cent.

Note that there are no changes in revenues to the ANSP. Departures and approaches are generally paid according to great circle distances between the aircraft's points of entry and exit in the area within which it is controlled, instead of to actual distance travelled or time under control. Therefore the ANSP has no incentive to carry out the investment, other than by increasing the asset base used for regulatory purposes.[8] Moreover, if ATM was remunerated through a share of airline ticket prices, and the airline markets in question were competitive, the investment may actually reduce ANSP revenues, as airlines pass fuel savings to passengers via lower air ticket prices. On the other hand, cash benefits to the airlines are sufficient to justify the investment. If the regulatory setting allowed it, the ANSP could negotiate with the airlines an increase in the air navigation charge to fund the project, leaving both airlines and ANSP better off.

This is a similar situation to that identified in the runway enlargement project example discussed in Chapter 4, section 6. The financial analysis concludes that the infrastructure operator does not have a financial incentive to carry out the infrastructure improvement (other than through inflating the regulatory asset base, should the operator be subject to economic regulation, as discussed in section 5, Chapter 4). But the economic analysis, by identifying who benefits from the project and by how much,

---

volumes with and without the project would differ. However, for this type of project the effects can be expected to be small, and omitting generated traffic would only make a small difference to the estimated returns.
8   See section 2 above for a discussion of pricing policy and investment incentives in ATM; and Chapter 4, section 5, for a discussion of the incentives to investment under economic regulation of charges.

enables the infrastructure operator and the main economic beneficiaries to negotiate sharing the benefits from, and the financing of, the project, in order to generate the incentive to carry out the investment. Note that the beneficiaries that would have an incentive to help finance the project include not only the airlines, which would save fuel with the project, but also residents in the vicinity of the airport, who would benefit from lower noise and air pollution. The latter group could be approached directly, through the airport or through their political representatives. However, normally the most economically efficient way to tackle the externality would be to internalise it. This could be done with a tax on the air ticket or on the landing charge that reflected the costs from noise and air pollution. That would devolve the issue of incentives to carry out the project back to the airlines and the ANSP. It would reinforce the incentive of the airlines to encourage the ANSP to carry out the project (should the ANSP not have it already in the form of an incentive to expand the regulatory asset base). The most economically efficient way would be (subject to second-best considerations) by means of proposing an increase in the approach and departure charge.

Chapter 6
Airlines

Introduction

When acquiring new aircraft, airlines alter their fleet along two dimensions: expansion and renewal. Airlines expand their fleets to address the growth in the demand for air travel and, if the airline is commercially successful, to capture market share from competitors. Fleet expansion can take place by buying more and/or bigger aircraft. Today airlines operate in highly competitive environments, reflected in thin profit margins and high operator turnover (namely entry and exit activity). This means that fleet expansion plans should be based more on the airline having a clear competitive advantage, enabling it to operate profitably, than merely on expected traffic growth, since a loss-making airline that grows its fleet can only expect to grow its losses, negating the investment case for a fleet expansion.

Likewise, the fleet replacement decision tends to be determined by commercial decisions under competitive conditions. Properly maintained, aircraft can fly for many decades. Some aircraft dating from the 1930s and 1940s are still airworthy today and still operate commercially, including, notably, the Douglas DC-3. And yet, airlines tend to replace their aircraft after around 20 years of operation, with some airlines doing so much earlier. The justification for this is twofold: first, new aircraft tend to offer significant operating efficiency improvements; and second, passengers may appreciate newer aircraft. Therefore competition incentivises airlines to renew their fleet despite the aircraft in their current fleet having many airworthy years ahead of them.

The decision-making process as to what aircraft to buy and when is purely a commercial one and follows the standard financial business plan. Airlines model their current and planned route network to see how different aircraft would perform in the various routes. They then set the operational suitability of different aircraft types against the price, delivery and after-sale service offered by the various manufacturers, reaching a decision on purely commercial grounds.[1]

Given that the decision to invest is a commercial one and that it is frequently made under competitive markets, economic appraisal, distinct from financial appraisal, has a limited role in the aircraft investment decision. Its main use would be in shedding light on effects on investment returns of changes in government policies in situations where there are market distortions. A topical case at the time of writing is the introduction of pricing mechanisms to internalise externalities, such as greenhouse gas (GHG) emissions. A financial analysis would capture the impact of the environmental measures proposed by authorities, and only that. Gauging the full extent of the distortion caused by environmental externalities and, hence, the possible impact of potential additional future policy changes, would require an economic appraisal.

Another example of the use of economic appraisal would be to evaluate air services that enjoy government subsidies, whether paid directly to airlines or indirectly to, say, the airports. Economic analysis would unveil the potential return of the air service should the government change its policy on subsidies to air transport in the future.

More generally, economic appraisal is useful when estimating the economic returns of air services to society, which is useful to airlines when attempting to influence the government policy-making process. The value of air transport to society is often claimed to be measured by the contribution of the air transport industry to gross domestic product (GDP). Since airlines tend to pay for all capital costs, including infrastructure, it would follow that the value of air transport to society could be measured by the contribution of airline services to GDP, minus non-internalised environmental costs. In reality, such measures greatly underestimate the economic contribution of aviation to

---

1 See Clark 2007 for a guide to the issues involved.

society, which has more to do with its poor substitutability, a condition that is poorly captured by GDP metrics as will be shown in this chapter.

It should be noted that in addition to the traditional role of economic appraisal – or cost-benefit analysis – in informing about capital allocation in investment appraisal, the techniques used also play a role in the investment process indirectly by helping forecast demand. The concepts of value of time, generalised cost of travel and schedule delay allow the airline to forecast demand for a new route. The case of expanding a runway in Chapter 4, section 6 shows how introducing a direct service produces time savings, which are valuable to passengers, giving an indication of the demand generation potential of the service.[2] However, since such analysis falls more closely into the realm of demand forecasting than of measuring investment returns, it is not pursued further here.

This chapter has two main objectives. First, it looks at estimating the returns of aircraft fleet investments in the presence of market distortions. One key distortion is external costs via environmental pollution. The chapter deals with this issue as generically as possible, without entering into discussion on alternative ways of internalising externalities, whether through taxes, emissions trading, or regulation. For simplicity, the examples use taxes to illustrate the effects of internalising external costs. The chapter starts by looking at fleet replacement and follows with fleet expansion, including the valuation of options on aircraft.

---

2 In the past, when airline capacity was subject to government regulation, general transport planning concepts like traffic diversion and generation would have been useful in the process of requesting new routings. Indeed the concept of frequency delay and other concepts, such as stochastic delay (passenger diversion through high load factors), were developed by airline analysts at the time. Today, the freedom to establish new routes, combined with the high mobility of aircraft, mean that formal demand modelling is less critical. Many airlines use proprietary information on demand flows and test route potential with some form of gravity model (see Doganis 2010 and Vasigh et al. 2008). Some airlines, though, are readier to simply test out potential through trial services. Airlines tentatively deploy aircraft on a new route and, depending on results, decide whether to keep, grow or withdraw a service.

The second objective of the chapter is to contrast correct and incorrect ways of estimating the socio-economic benefits of aviation to society. In so doing the chapter also illustrates investment appraisal when the alternative to the project is another transport mode, or when scenario-building focuses on inter-modal competition.

1 Fleet Replacement

An airline has a short- to medium-range fleet of 50 Code-C aircraft ('Old C'), with a seating density of 150 seats. This fleet is approaching 20 years of age and the airline is considering replacing it with newer aircraft of the same category ('New C'). The New C aircraft would cost €45 million each and would be delivered over three years. As deliveries are made, the Old C aircraft would be sold at €5 million each. At the average sector length of 1,000 kilometres and load factor of 70 per cent, which characterise the network and operations of the airline, New C will be 15 per cent more cost-efficient than Old C, reducing unit costs from 6 cents per available seat-kilometre (ASK) to 5.1 cents per ASK. At these sector length and load factor, New C would also reduce GHG emissions by 17 per cent, or from 0.33 kg/RPK to 0.274 kg/RPK, where RPK stands for revenue passenger-kilometre.

Each aircraft in the fleet operates an average of 229.95 million RPK each year, and the competitive position of the airline is such that financially each aircraft produces an operating margin before depreciation and airline overheads of 25 per cent on sales. The operating margin of the airline as a whole would be lower after including costs for administration, marketing, etc. Since such overheads are assumed to be the same with and without the aircraft replacement, they are ignored.

The management of the airline is fairly certain that the company can maintain its competitive position over the medium to long term, so it will be able to sustain the current degree of pricing power in the future. The main benefit of the project will therefore consist of increased operating profits through cost savings. In addition, there is uncertainty about

Airlines

possible future taxes for carbon emissions. There has been talk of them being introduced, and although the authorities are not expected to reach a decision before the airline needs to decide on the project, even if they decide against their introduction, the possibility of new policy initiatives for such taxes would remain through the operating life of the new aircraft. Current projections are that the marginal cost of carbon and, hence, the possible extent of such a tax, would average €40 per tonne over the long term.

For simplicity, other taxes on revenues or costs are excluded from the analysis. In any case, since revenues with and without the project would be the same, the effect of taxes on revenue can be ignored in the economic calculation. The effects of including taxes on costs would depend on the nature of the taxes. Assuming that they are proportional to costs, the effect would be equivalent to the effect of a tax on GHG emissions, illustrated below. Finally, it is assumed that the newly acquired aircraft would have the same residual value after 20 years of operation as the Old-C aircraft would command during project implementation.

Table 6.1 shows the results of the investment appraisal, assuming that GHG emissions are not taxed. The calculation of financial returns consists simply of comparing the operating profits (cash generated from operations) that the airline will generate with the New C aircraft (row 6) against those that it would generate with the current Old C fleet (row 13), subtracting the investment cost (row 16) and adding the proceeds from the sale of the current fleet (row 17). The resulting net financial flows (row 18) add to a net present value of €71 million, equivalent to a financial return of 5.5 per cent. The estimate is conservative as it does not include difficult to quantify factors such as some passengers being warded off by old-looking aircraft. Still, the return is not large and is close to the cost of capital of 5 per cent.

An economic appraisal of the investment would consist of adding the external costs with and without the project (rows 7 and 14, respectively) to the flows used for the financial appraisal. The result (row 19) shows that the economic return of the project, at 7.7 per cent, is higher than the financial return as the newer aircraft produce fewer emissions than the older aircraft. This higher return tells management that under conditions of efficient pricing, that is, under conditions where externalities are internalised, the project would be more profitable than with prices resulting from current government policy.

Table 6.2 includes the same calculation assuming that GHG emissions are fully internalised through a tax on fuel. The response of all competing airlines would be to pass on the cost of the tax to users by increasing airline ticket prices, in order to preserve normal profitability levels characteristic of competitive markets. This is reflected in higher revenues (rows 4 and 11) and costs (rows 5 and 12). In turn, the external costs disappear (rows 7 and 14).

Such behaviour by airlines is akin to saying that air ticket prices are set by the market. In such circumstances, as far as the project airline is concerned, assuming that competing airlines renew their fleet, the yields (i.e. average revenue per passenger) faced by the airline will be the same with or without the project. In the 'without project' scenario, which assumes that the airline keeps the more polluting Old C aircraft, the airline will endure a higher tax bill than its competitors. This would mean that in the 'without project' scenario the profit margin of the airline when there is a GHG tax will be squeezed more than in the situation where there is no GHG tax, as seen in rows 13 of tables 6.1 and 6.2.

The result is that with the internalising of external GHG emissions costs, the financial return of the project would be higher, at 7.5 per cent instead of the 5.5 per cent achieved in the scenario with externalities. Also note that internalising externalities means that, other things remaining constant, the financial return and the economic return are equal. This result highlights the role that the economic return assumed in the case with no internalisation of externalities (Table 6.1) of signalling to the airline management the underlying desirability of the investment.

Table 6.1 Financial and economic returns on a fleet replacement project with external emissions costs

			Year \ PV	1	2	3	4	5	10	15	20	21	22
	With project												
(1)	ASK	(million)		16,425	16,425	16,425	16,425	16,425	16,425	16,425	16,425	16,425	16,425
(2)	RPK	(million)		11,498	11,498	11,498	11,498	11,498	11,498	11,498	11,498	11,498	11,498
(3)	Passengers	(million)		11.5	11.5	11.5	11.5	11.5	11.5	11.5	11.5	11.5	11.5
(4)	Revenues	(EUR m)	14,702	1,117	1,117	1,117	1,117	1,117	1,117	1,117	1,117	1,117	1,117
(5)	Operating costs	(EUR m)	11,193	956	897	838	838	838	838	838	838	838	838
(6) = (4) − (5)	Profits	(EUR m)	3,509	161	220	279	279	279	279	279	279	279	279
(7)	Emissions costs	(EUR m)	1,658	126	126	126	126	126	126	126	126	126	126
	Without project												
(8)	ASK	(million)		16,425	16,425	16,425	16,425	16,425	16,425	16,425	16,425	16,425	16,425
(9)	RPK	(million)		11,498	11,498	11,498	11,498	11,498	11,498	11,498	11,498	11,498	11,498
(10)	Passengers	(million)		11.5	11.5	11.5	11.5	11.5	11.5	11.5	11.5	11.5	11.5
(11)	Revenues	(EUR m)	14,702	1,117	1,117	1,117	1,117	1,117	1,117	1,117	1,117	1,117	1,117
(12)	Operating costs	(EUR m)	12,972	986	986	986	986	986	986	986	986	986	986
(13) = (11) − (12)	Profits	(EUR m)	1,730	131	131	131	131	131	131	131	131	131	131
(14)	Emissions costs	(EUR m)	1,998	152	152	152	152	152	152	152	152	152	152
	Net project flows												
(15)	Aircraft deliveries	(units)		10	20	20							
(16)	Investment cost	(EUR m)	2,022	450	900	900							
(17)	Sale of old aircraft	(EUR m)	419	50	100	100					50	100	100
(18) = (6) − (13) − (16) + (17)	Financial flows	(EUR m)	71	−370	−711	−652	148	148	148	148	198	248	248
	FRR		5.5%										
(19) = (18) − (7) + (14)	Economic flows	(EUR m)	410	−345	−686	−626	174	174	174	174	224	274	274
	ERR		7.7%										

Table 6.2 Financial and economic returns on a fleet replacement project with emissions costs internalised

		Year\PV	1	2	3	4	5	10	15	20	21	22
	With project											
(1)	ASK (million)		16,425	16,425	16,425	16,425	16,425	16,425	16,425	16,425	16,425	16,425
(2)	RPK (million)		11,498	11,498	11,498	11,498	11,498	11,498	11,498	11,498	11,498	11,498
(3)	Passengers (million)		11.5	11.5	11.5	11.5	11.5	11.5	11.5	11.5	11.5	11.5
(4)	Revenues (EUR m)	16,360	1,243	1,243	1,243	1,243	1,243	1,243	1,243	1,243	1,243	1,243
(5)	Operating costs (EUR m)	12,880	1,103	1,033	964	964	964	964	964	964	964	964
(6) = (4) − (5)	Profits (EUR m)	3,480	140	210	279	279	279	279	279	279	279	279
(7)	Emissions costs (EUR m)	0	0	0	0	0	0	0	0	0	0	0
	Without project											
(8)	ASK (million)		16,425	16,425	16,425	16,425	16,425	16,425	16,425	16,425	16,425	16,425
(9)	RPK (million)		11,498	11,498	11,498	11,498	11,498	11,498	11,498	11,498	11,498	11,498
(10)	Passengers (million)		11.5	11.5	11.5	11.5	11.5	11.5	11.5	11.5	11.5	11.5
(11)	Revenues (EUR m)	16,360	1,243	1,243	1,243	1,243	1,243	1,243	1,243	1,243	1,243	1,243
(12)	Operating costs (EUR m)	14,970	1,137	1,137	1,137	1,137	1,137	1,137	1,137	1,137	1,137	1,137
(13) = (11) − (12)	Profits (EUR m)	1,390	106	106	106	106	106	106	106	106	106	106
(14)	Emissions costs (EUR m)	0	0	0	0	0	0	0	0	0	0	0
	Net project flows											
(15)	Aircraft deliveries (units)		10	20	20							
(16)	Investment cost (EUR m)	2,022	450	900	900							
(17)	Sale of old aircraft (EUR m)	419	50	100	100					50	100	100
(18)=(6) −(13) −(16) + (17)	Financial flows (EUR m)	381	−365	−696	−626	174	174	174	174	224	274	274
	FRR	7.5%										
(19) = (18) − (7) + (14)	Economic flows (EUR m)	381	−365	−696	−626	174	174	174	174	224	274	274
	ERR	7.5%										

It is worth mentioning in passing that the example above also illustrates how, despite aircraft being airworthy for many decades, competition encourages more frequent fleet renewal so long as aircraft manufacturers bring to the market aircraft with improved operating cost efficiency. In tandem, aircraft manufacturers are incentivised to do so through competition to sell aircraft. It also illustrates how internalising air transport externalities strengthens those incentives, as the penalty for not renewing the fleet increases (a foregone return of 7.7 per cent with internalisation instead of 5.5 per cent without internalisation).

The case chosen here illustrates pricing distortions through emissions of GHG. But estimates of economic returns retain the same signalling power to airline management about any other price distortion, including subsidies, import levies, taxes on labour, etc. The economic return signals the real return of a project, in terms of resources used and produced, and sheds light on risks regarding future government policy.

2 Fleet Expansion

The aircraft acquisition programmes of airlines generally include fleet renewal and expansion simultaneously. The investment appraisal of the fleet renewal and fleet expansion components differ in two respects. The first is that, in competitive conditions, the analysis of a fleet expansion project does not require an ad hoc 'without project' scenario to be devised. Should the airline not expand its fleet, the traffic is simply absorbed by competing airlines. That is, under competitive conditions the 'without project' scenario is simply the opportunity cost of the project inputs. The financial evaluation of the investment follows a standard commercial business plan, where the airline sets expected revenues against investment and operating costs, instead of the differential cash flow approach applied for fleet replacement. The economic returns will coincide with the financial return, other than for the usual corrections for taxes and externalities.

If the airline market is not competitive, a counterfactual or 'without project' scenario is necessary, in order to conceive what passengers would do in the absence of additional air transport capacity. In the financial analysis, the airline, facing growing demand, will be capable of increasing prices and will exert monopoly profits. The economic analysis will register the inefficiencies of doing this. Incidentally, measuring the economic returns of airline fleet expansions in conditions of monopoly is illustrative of the economic value of air transport, which is explored in the next section.

The second difference between the analysis of fleet replacement and expansion is that, generally, an expansion involves a greater degree of risk, in two respects, including the amount of future demand growth and the extent of future competition. Whereas demand for air travel tends to grow over the long term, the degree of growth depends on general economic growth, which is less certain. Moreover, economic growth and demand are cyclical, and aircraft deliveries may coincide with traffic downturns. As for the degree of competition, when expanding its fleet the airline will be generally venturing into new, lesser-known markets. This may involve entering into competition with airlines with which the project airline was hitherto not competing and which managers of the project airline may know less well. The result is that airlines will tend to have less visibility of future competitive conditions and may therefore wish to have greater flexibility to decide on the extent of the capacity expansion. For these reasons, airline fleet expansion programmes generally combine firm aircraft orders with options.

2.1   Firm Orders and Options

An option to buy an aircraft is a right, but not an obligation, to buy an aircraft in the future. Given that options may cost money, or at least will be contingent upon placing firm orders, the question then becomes: How much are those options worth to the airline and, therefore, how much should it be willing to pay for investing in them?

Table 6.3 Traffic and investment return scenarios for an airline considering an expansion of its fleet

		Year\PV	1	2	3	4	5	10	15	20	24
(1)	RPK	(million)	11,498	11,842	12,198	12,564	12,941	15,002	17,391	20,161	22,691
(2)	Diff. RPK	(million)		345	700	1,066	1,443	3,504	5,894	8,663	11,194
(3)	Required fleet	(units)		2	3	5	6	15	26	38	49
(4)	Aircraft options exercised	(units)				6					
(6)	Expanded RPK	(million)				1,066	1,380	1,380	1,380	1,380	1,380
(7)	Investment	(EUR m)	257			270					
	Optimistic scenario										
(8)	Operating cash flow	(EUR m)	374	0	0	26	34	34	34	34	64
(9) = (8) – (7)	Net flows	(EUR m)	152	0	0	−244	34	34	34	34	64
	FRR		12.6%								
	Base scenario										
(10)	Operating cash flow	(EUR m)	228	0	0	16	20	20	20	20	50
(11)=(10) –(7)	Net flows	(EUR m)	6	0	0	−254	20	20	20	20	50
	FRR		5.3%								
	Pessimistic scenario										
(12)	Operating cash flow	(EUR m)	82	0	0	5	7	7	7	7	37
(13)=(12) –(7)	Net flows	(EUR m)	−140	0	0	−265	7	7	7	7	37
	FRR		N/A								

The value of buying options instead of firm orders depends on the potential future payoffs and the degree of uncertainty surrounding those payoffs. Continuing with the airline example in the previous section, let us assume that airline management is fairly certain about future growth prospects. Traffic is expected to grow at 3 per cent per year and infrastructure constraints combined with sound economic growth means that they can expect the expansion to take place without affecting real yields (that is, yields net of inflation).

Row 1 in Table 6.3 describes the total RPKs that would result from growing existing traffic by 3 per cent per year. Row 2 includes the increase in RPKs relative to the starting year, and row 3 the number of additional New C aircraft that would be required to accommodate such traffic levels.

Assume that assemblage capacity constraints mean that in the first three years of the project the aircraft manufacturer can only deliver the 50 firm orders already placed for fleet replacement. The airline will therefore have to meet that growing demand by delaying the phasing out of Old C aircraft. But in year 4 the manufacturer has free slots to deliver six aircraft. Taking delivery of six additional aircraft in year 4 would generate 1,066 million RPKs in year 4 and 1,380 million thereafter (row 6). The question management faces is whether it is worth ordering those six new aircraft.

Suppose that he Optimistic scenario reflects the conditions expected by management. In this scenario, the present value of the operating cash flow generated from the operation of the six aircraft will be €374 million (row 8). At an aircraft price of €45 million per unit, the investment has a present value of €257 million (row 7), which at a discount rate of 5 per cent means that the investment will be worth €152 million (row 9), or generate a return of 12.6 per cent. The airline deems such a return adequate and will place that firm order for aircraft. Indeed, if the airline is quite certain about future prospects, the airline will go beyond those six orders, as similar analysis of further deliveries in the future will show also positive returns. Following the demand projections in row 2, by year 15, for example, it will require 26 new aircraft.

But let us assume instead that future prospects are far less certain and positive. In particular, airline management is divided about future competition prospects. Pessimists argue that the airline market will turn increasingly competitive, particularly because of the growth of low-cost carriers, forcing the airline to decrease real yields by as much as 20 per cent. The outcome of such a fall in yields would be as depicted by the Pessimistic scenario in Table 6.3. Buying the six aircraft will only generate €82 million of operating profits (row 12) which, when set against the cost of investing in the six aircraft, will mean a negative value of the investment of €140 million (row 13), and a negative return on investment.

Management agreed that the Optimistic and Pessimistic scenarios are within the realm of the possible, and took them as extreme cases. Other managers saw less extreme scenarios and, taking together the opinion of all managers, they built an additional scenario referred to as the 'Base case', whereby real yields would fall by 10 per cent. Under this scenario, the investment in six aircraft would produce operating profits with a present value of €228 million. However, when setting this against the investment cost, the project would have a present value of only €6 million, or a return of 5.3 per cent, deemed borderline by management and probably not worth the risk.

Such a result, however, is independent of the dispersion of possible positive and negative outcomes. The net present value (NPV) of the investment is equally a gain of €6 million, whether management considers that this base scenario constitutes a certain outcome, whether the base case can only be given a probability of 50 per cent and the two extreme scenarios of 25 per cent each, or indeed any other probabilities resulting from different probability distribution of outcomes.[3] And yet it is clear that the greater the dispersion of possible results, the greater

---

3 This exercise will be solved calculating the option value through the Black–Scholes method, therefore the underlying probability distribution of outcomes is assumed to be lognormal. Other methods of calculating option value can relax this assumption. An example of such an alternative method, the binomial method, is illustrated in Chapter 7, section 2, regarding the option value of research and development in the aeronautics industry.

the likelihood that the Optimistic scenario will materialise. If managers decided not to invest and in the following years the market were to evolve in a way that vindicated the view of the optimists, the airline would have foregone a profitable investment opportunity. Options enable the airline to delay taking a decision on whether to acquire the aircraft until future market trends become clearer, reducing the risks involved in placing the order while enabling them to profit from the investment opportunity should markets evolve favourably. In fact, options are most useful in circumstances when the present value of the project is not satisfactory (after all, if it was satisfactory the airline would simply place firm orders) but there is a reasonable chance that circumstances might evolve in the future in such a way that the project would offer a satisfactory return.

Such an option must obviously be valuable to the airline. The question then becomes how much the options are worth to the airline and, therefore, what would be the maximum price that the airline should be willing to pay for them.

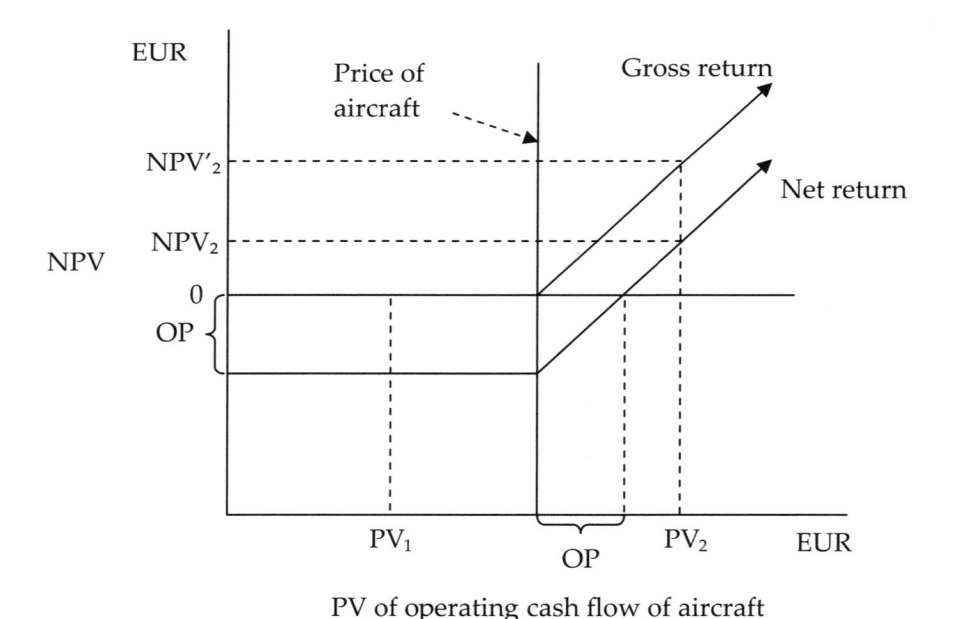

Figure 6.1 **Option price and value at expiration and investment returns**

2.2 The Value of Options on Aircraft

The relationship between option value, option price and project profitability is illustrated in Figure 6.1, which summarises the situation the airline faces when exercising the option a few years into the future.[4] The vertical axis of Figure 6.1 represents the NPV of the project per aircraft, calculated by comparing the cash flow from operating the aircraft estimated at the time of exercising the option, the price of the aircraft (the exercise price of the option) and the price at which the option was bought. The horizontal axis measures the present value (PV) of operating the aircraft, estimated at the time of exercising the option.

OP is the option price, or price paid by the airline for the option at the time it was bought, inflated by the cost of capital and inflation. PV_1 could represent either the Pessimistic or Base scenarios materialising. In any of those two cases, the airline will not exercise the option, which will expire and lose its value. The investment in the options would have resulted in a loss to the airline equal to the present value of the money paid for the option, measured by OP on the vertical axis.

However, if at the time of exercising the option it turns out that trends in the airline market look favourable and management views shift to those projected in more optimistic scenarios, so that a discounted cash flow (DCF) calculation produces a positive NPV, the option would be exercised. The 'Gross return' line measures the NPV calculated from the aircraft purchase cost (which constitutes the exercise price of the option) and the expected cash flows from operating the aircraft, ignoring the present value of the option price – that is, taking bygones as bygones. The 'Net return' is the gross return minus the present value of the price paid for the option. Assuming that at the time of exercising the option the estimated operating profit from operating the aircraft is PV_2, then buying the aircraft would produce a return of NPV'_2. That will also be the value of the option at the time. The option would clearly be exercised. Even after subtracting the present

---

[4] The introduction to option valuation here is schematic as such material is broadly available in the literature. Accessible sources include Kodukula and Papudesu 2006 and Brealey et al 2008.

value of the money paid for the option, an investment decision to first buy options and then exercise them would still have made sense, with a positive return of NPV_2 on the vertical axis.

An interesting case would be one where the estimated PV of the future cash flows of operating the aircraft was positive but less than the present value of the price paid for the options, that is, if the PV on the horizontal axis fell somewhere within the OP bracket on the horizontal axis. In such a case, the net return of the project, including the price paid for the option in the past, would be negative. However, in investment decisions what matters is the return expected at the point when the decision is made – bygones are bygones. At that point, looking forward, the airline can expect to make a positive return – above the cost of capital – by exercising the option, as determined by the gross return schedule, and should therefore exercise the option, even though the value of the option is less than the present value of the price paid for it.

The illustration has focused on what the airline should do at the time of making the decision of whether exercising the option or not given the price and value of the option. However, this is a situation the airline will face a few years into the future. The question that the airline has to address is what the value of each option is when deciding whether to buy the options, a few years ahead of deciding whether to exercise them.

The standard method of calculating option value is the Black–Scholes formula, suitable for financial options with a predetermined exercise date (called 'European options').[5] The expression is as follows:[6]

---

5 Alternatively, American options can be exercised at any time before the expiry date. Whereas airlines are normally free to convert options into fixed orders at any time (the formal exercise of the option), the actual exercise date of the option (the delivery of the aircraft) is constrained by assembly line schedules, effectively removing exercise date flexibility from a standard American option. Therefore in practice aircraft options tend to have elements of European rather than American options. Should the aircraft manufacturer offer sufficient flexibility regarding delivery dates, the options could then be considered American. The modelling of the actual option facing an airline must be tailored to the circumstances applying to each case. The important thing here is to illustrate the use of real option analysis to help make investment decisions. Chapter 7, section 2.1, includes an example of valuing a real option using the binomial method, which is better suited to American options.

6 An exposition of the theory of real options or the theoretical justification

$$C = N(d_1)S - N(d_2)Ke^{-rT}$$

... where C is the option value, S is the value of the underlying asset, or the DCF of operating the aircraft, K is the strike price of the option, or the cost of the firm aircraft orders, r is the risk-free rate of return and T is the time to maturity of the option, N is the standard normal distribution and d_1 and d_2 are option parameters as follows:

$$d_1 = \frac{\ln\left(\frac{S}{K}\right) + \left(r + \frac{\sigma^2}{2}\right)T}{\sigma\sqrt{T}}$$

$$d_2 = d_1 - \sigma\sqrt{T}$$

... where σ is the volatility of the cash flows of the underlying asset, that is, of operating the aircraft, which can be estimated as follows:[7]

$$\sigma = \frac{\ln\left(\frac{S_{opt}}{S_{pes}}\right)}{4\sqrt{t}}$$

---

behind the Black–Scholes method is beyond the scope of this book; the reader should consult the many available references. For a formal exposition of the case for real option analysis see Dixit and Pyndick 1994. For a more accessible guide to real option applications see Kodukula and Papudesu 2006. Koller et al. (2010) also include accessible applications using alternative procedures.

7   This is just one method of calculating volatility, based on management assumptions about future scenarios. Other methods of estimating volatility rely either on extensive historical data or on assumptions by the analyst. Alternatively, volatility can be borrowed from projects or securities that could be expected to have similar cash flow profiles and are subject to similar degrees of uncertainty as the project being appraised. See Kodukula and Papudesu 2006 for accessible discussion of volatility estimation, and Koller et al. 2010 for a worked example using a traded security as a proxy.

... where S_{opt} is the underlying asset value under the optimistic scenario, S_{pes} is the underlying asset value under the pessimistic scenario, and t is the project lifetime.

The calculation process therefore starts with an estimate of the volatility of returns which, in our example at hand, as detailed in Table 6.3, would be estimated as follows:

$$\sigma = \frac{\ln\left(\frac{374}{82}\right)}{4\sqrt{20}} = 8.47 \, per \, cent$$

Given the maximum (optimistic) return of €374 million, and a minimum (pessimistic) return of €82 million, estimated over a project life of 20 years, the volatility of the underlying asset class, that is of the cash flows of operating the aircraft, is 8.5 per cent. With this the option parameter d_1 can be estimated as follows:

$$d_1 = \frac{\ln\left(\frac{228}{270}\right) + \left(0.05 + \frac{0.847^2}{2}\right)4}{0.847\sqrt{4}} = -0.6132$$

And with the value of d_1 the parameter d_2 is calculated as follows:

$$d_2 = -0.6132 - 0.847\sqrt{4} = -0.7825$$

The formula of the value of the option would then be:

$$C = N(-0.6132)228 - N(-0.7825)270e^{-0.05 \times 4}$$

The $N(d_1)$ and $N(d_2)$ functions are standard normal distributions, which normally come as default functions in spreadsheets. The resulting figures are:

$$N(-0.6132) = 0.2699$$

$$N(-0.7825) = 0.2170$$

The resulting value of the options is therefore:

$$C = (0.2699 \times 228) - (0.2170 \times 270)e^{-0.05 \times 4} = 13.6281$$

The value of the options for the six aircraft included in our example adds up to €13.6 million, which works out to an option value of almost €2.3 million per aircraft. Therefore, even though under the base case the present value of the operating cash flows for the six aircraft is very low, at €6 million, it still pays the airline to buy options for the six aircraft, so long as those options cost less than €2.3 million each.

The calculation of the option value in this section of the chapter has consisted purely of a financial value. Options can also be calculated for economic values. This is illustrated in Chapter 7, section 2.2, where investments in the aeronautical sector are discussed.

3 The Value of Air Transport

Air transport pays for itself, both in the passenger and the freight sectors. Aviation is one of the few modes of transport that covers all operating and infrastructure costs. The one exception at the time of writing is emissions costs. But air transport could pay for all its emissions costs and remain viable, and strongly so. Instead, passenger rail transport tends to rely on subsidies, whereas rail freight has a greater ability to pay for itself. Road transport rarely pays for the cost of infrastructure, although it is generally accepted that it is a transport mode that could pay for itself. Maritime transport, particularly maritime freight, also tends to pay for all costs except emissions, a situation similar to air transport. Maritime passenger transport is viable only on a relatively small number of routes.

The financial and economic viability of air transport arises from a substantial competitive advantage relative to other modes, based on the ability to provide fast and safe transportation along longer distances at an affordable cost. The economic viability of air transport also reflects the value of air transport to society. This section deals with how to measure such value.

3.1 Invalid Approaches to Measuring Value

It is at times implied that the value of air transport to society is measured by an estimation of airline profitability, assuming that airlines pay for all its environmental costs. This approach is mistaken for three main reasons. First and most importantly, it does not take into account additional value to consumers in the form of consumer surplus. Second, it ignores the producer surplus of infrastructure providers. Whereas the opportunity cost of capital invested in infrastructure should be reflected in airline operating costs and, in principle, should not require any additional treatment, most other passenger transport modes do not pay for their infrastructure costs, hence it is incorrect to treat infrastructure costs for aviation and for other modes of transport equally. Third, it does not take into account subsidies to operators of alternative modes of transport.

Another attempt to measure the value of aviation to society would be by measuring the contribution of aviation to Gross Domestic Product (GDP). To recap, GDP consists of any of three equivalent measures, namely, the monetary value of output produced, the total income received by the owners of the factors of production, or the net expenditure on the sector.[8] However, GDP does not include any measure of consumer surplus. Moreover it does not consider the opportunity cost of resources and, therefore, whether the output should be produced at all. After all, an unviable business may still generate GDP.[9] So long

---

[8] For accessible presentations of the components of GDP see Moss 2007 or Johnson and Briscoe 1995.
[9] Imagine a company that makes no profit – and that never will because there are better technologies around – and pays €1 million in salaries a year.

Airlines 179

as the salaries paid by a company are higher than the financial losses of the company in absolute terms, the company will make a positive contribution to GDP. Finally, GDP does not measure environmental externalities.[10]

To illustrate this it is worth first pinning down the differences in gauging air transport operations in terms of contribution to GDP and in terms of generalised costs. These are illustrated in Table 6.4, which includes for a hypothetical passenger comparisons of contribution to GDP of a trip by air, and the generalised cost of the same trip, across different route lengths. The figures are indicative since the purpose of the exercise is to illustrate the measuring processes rather than to arrive at any empirical finding.

Rows 1 to 4 include the contribution of the trip to GDP, measured through income, including income to workers in the air transport sector via salaries, income to the owners of capital via profit to the airline and other service providers, and income to the government via taxes. For a route of 500 kilometres the contribution of the trip to GDP would be €75, for a trip of 1,000 kilometres it would be €125, and for a trip of 2,000 kilometres it would be €150 (row 5).

The generalised cost of travel is calculated for three transport modes: air, high-speed rail (HSR), and road via private car. It is assumed that in the case of air transport the trip involves 1.5 hours of access and egress time, both including passenger processing time at the terminal. This is lower for rail, as train stations tend

---

It will contribute to GDP by €1 million, by means of income to labour. However, once it is recognised that labour and capital have an opportunity cost, that is, viable alternative uses, then salaries are not a benefit, but a cost, and the capital invested will have to be charged an opportunity cost of capital – the rate of discount on an investment appraisal. In those circumstances, even though the company still makes a contribution to GDP, the negative return indicates that its resources would be better deployed on some other activity.

10 As argued in Chapter 2, section 7.1, in perfectly competitive markets the observed financial profitability of a project can be taken as the economic profitability. Varian 1992 shows how under such perfect conditions, income and economic viability coincide. However, it should be borne in mind that income in that context does not correspond to the GDP measure, which does not allow for the opportunity cost of factors of production.

to be closer to city centres than airports and involve shorter passenger processing time. In-vehicle time varies with route length. The summation of these three time components constitutes total (door to door) travel time. Air travel performs better in terms of travel time relative to other transport modes as route distance increases. It is assumed that the value of time to the passenger is €30 per hour. Since the comparison is for the same passenger, the value of time is taken to be the same for all transport modes. The product of total travel time and value of time yields the time cost component of the generalised cost (row 11).

The money cost of travel includes all expenditures by society to operate the service (rows 12 to 15). In the case of air, all costs are included in the ticket price. In the case of HSR the ticket price includes only operating cost, but no infrastructure cost, which is included as an additional cost, borne by the taxpayer. For road, all operating costs are paid directly by the user, except for infrastructure costs. Finally, the generalised cost includes all external costs to members of society other than the transport user and the producer of transport services, including emissions of GHG, air pollutants and noise (rows 16 to 19). It is assumed that HSR is powered fully by renewable energy, yielding no GHG or pollutant emissions. For simplicity, the calculation ignores safety issues. Including them would favour air transport, particularly against road transport.

The generalised cost consists of the summation of the time, money and external costs. A difference is made between generalised cost for the user – or behavioural generalised cost – and for society at large – or total generalised cost. The former includes only costs borne by the user (row 20), whereas the latter includes those borne by the user and by other members of society (row 21). For each route length, the lowest generalised cost among the various modes (i.e. the best option) is circled, and the best alternative to air transport is underlined. The value created by aviation is calculated by comparing the generalised cost of aviation to that of the best alternative transport mode. Again, a different value is calculated for the user (row 22) and for society at large (row 23). The results show that aviation creates value in longer distances. This does not mean that it cannot create value in shorter distances, but it would tend to do so only on routes where for reasons of say, geography or low traffic density, there is poor provision of alternative modes of transport. Also, the example of 1,000 kilometres, where the best alternative to the user differs from the best alternative to society, illustrates how inefficient pricing or subsidies (in the current case largely the latter) could shift traffic to less socially efficient modes.

Comparing the contribution to GDP (row 5) and the generalised cost (row 21) reveals how GDP misses out on signalling whether production is worthwhile. For shorter distances, whereas aviation may make a contribution to GDP, it may well be that society may be better off investing in alternative modes of transport. For longer distances, contribution to GDP grossly underestimates the value of aviation since it excludes non-monetised benefits. All in all, the fault of contribution to GDP as a guide to investment decisions lies in the fact that it does not correct for price distortions and ignores opportunity costs.

One may be tempted to construct a measure of the full value to society of the output produced by aviation by adding its contribution to income plus the savings in generalised cost – the latter effectively being the consumer surplus attributable to aviation net of other resources invested by third parties. The resulting 'hybrid' measure (row 24) would bring the GDP figure closer to opportunity cost. So in the illustration of travel on a 500-kilometre route, the value of the output of aviation would fall from €75 per passenger to €1, reflecting the fact that other activities would produce higher income. At the other extreme, the value figure would increase from €150, as measured by GDP, to €535 after taking into account all of the consumer surplus and other resources used.

However, such a hybrid measure of value can only be considered a curiosity of no valid practical use. It does not measure income as it takes into account non-money flows, particularly time savings. Likewise, it cannot guide investment decisions because it does not measure correctly resource costs (most importantly, GDP computes labour costs as a benefit) and incorrectly double-counts tax revenues as a benefit.

Table 6.4 Contribution to GDP and generalised cost of a hypothetical passenger trip across various route lengths

	Route length (km)		500			1,000			2,000	
			Air	HS Rail	Road	Air	HS Rail	Road	Air	HS Rail	Road
	CONTRIBUTION TO GDP										
(1)	Salaries	(EUR)	30			50			60		
(2)	Airline profit	(EUR)	7.5			12.5			15		
(3)	Profit of service providers	(EUR)	15			25			30		
(4)	Taxes	(EUR)	22.5			37.5			45		
(5) = (1) + (2) + (3) + (4)		(EUR)	**75**			**125**			**150**		
	GENERALISED COST										
	Time cost										
(6)	Access time	(hours)	1.5	0.5	0	1.5	0.5	0	1.5	0.5	0
(7)	Egress time	(hours)	1.5	0.5	0	1.5	0.5	0	1.5	0.5	0
(8)	In-vehicle time	(hours)	0.75	2	4.5	1	4	9	2	9	18
(9) = (6) + (7) + (8)	Total travel time	(hours)	3.75	3	4.5	4	5	9	5	10	18
(10)	Value of time per hour	(EUR)	30	30	30	30	30	30	30	30	30
(11) = (9) × (10)	Time costs	(EUR)	112.5	90	135	120	150	270	150	300	540
	Money cost										
(12)	Ticket price	(EUR)	150	100	0	250	200	0	300	300	0

(13)	Other operating costs	(EUR)	0	0	0	0	0	0	0	250
(14)	Other infrastructure costs	(EUR)	0	50	50	0	100	100	200	60
(15) = (12) + (13) + (14)	Money cost of operation	(EUR)	150	150	65	250	300	125	300	500	310
	External costs										
(16)	GHG emissions	(EUR)	10	0	2	15	0	4	25	0	8
(17)	Air pollution	(EUR)	2	0	0.5	3	0	1	5	0	2
(18)	Noise	(EUR)	5	3	3	5	5	5	5	10	10
(19) = (16) + (17) + (18)	Total external cost	(EUR)	17	3	5.5	23	5	10	35	10	20
	Generalised cost										
(20) = (11) + (12) + (13)	User	(EUR)	262.5	190	<u>185</u>	370	<u>350</u>	370	<u>450</u>	<u>600</u>	790
(21) = (20) + (14) + (19)… …= (11) + (15) + (19)	Total	(EUR)	279.5	243	<u>205.5</u>	393	455	<u>405</u>	<u>485</u>	<u>810</u>	870
	Value of aviation										
(22) = (20) − (<u>20</u>)	User	(EUR)	−77.5			−20			150		
(23) = (21) − (<u>21</u>)	Total	(EUR)	**−74**			**12**			**385**		
(24) = (5) + (23)	Hybrid measure	(EUR)	1			137			535		

Note: lowest GC circled and best alternative to air transport underlined.

3.2 Valid Approach to Measuring Value

Instead, the viability of air transport should be measured using the same tools as are used in the economic appraisal of aviation investments presented so far, based on comparing generalised costs to society of alternative transport modes. This is because such an approach measures total welfare created to society, namely net willingness to pay for the output produced, regardless of whether it is actually paid or not, while valuing resources used in production at opportunity cost.

The estimation of the full value of aviation is illustrated by measuring the benefits generated to society by investment in an aircraft. It is important to emphasise that, while the input numbers are realistic, they do not refer to an actually existing route, and that the exercise consists of an illustration of the method of measurement, rather than producing an empirical finding. The illustration takes the same aircraft operation as used so far in Chapter 6 on airlines: a New-C code aircraft flying back and forth along a route of 1,000 kilometres at 70 per cent load factor. The aircraft will fly almost 230 million RPKs each year. Average GHG per passenger will be €14, based on an average cost of carbon of €40 per tonne. It is assumed that the airports involved in the route are close to urban areas, yielding a relatively high noise impact of €10 per passenger.[11] Likewise, the impact of air pollution will be relatively high at €5 per passenger. It is assumed that neither the emissions of GHG, air pollutants or noise are internalised. Also, it is assumed that the door-to-door trip by air would take four hours.

The alternative mode is HSR and it is assumed that 100 per cent of the electricity consumed is renewable or nuclear, so there are no emissions of GHG or air pollutants. It is also assumed that the train follows a thinly populated route, so

---

11 As a comparison, using original data from the UK government, Eurocontrol (2009) estimates that for the average airport the marginal noise cost per landing or take-off of the B-737-400, a standard type-C aircraft, is €67. This implies that the cost is about €134 per flight or about €1 per passenger per flight. CE Delft (2002) estimates the cost at between €2 and €5 depending on aircraft technology.

the noise impact is half that of aircraft. As is common with existing HSR services, at similar ticket prices to those offered by air, the rail service would require subsidies to cover costs. It is estimated conservatively that the proportion of costs not covered through operating revenues represents 25 per cent of total costs, including infrastructure costs. It is also assumed that the rail service is already operational, so that no investment is needed. Therefore, the exercise formally addresses the question of whether it pays to invest in an aircraft to cover the route, given that there is an HSR service already in operation.[12]

The door-to-door rail trip would take five hours, so that by travelling by air over the 1,000 kilometres, air travellers would save one hour. Assuming that the prices of the air and rail ticket are the same at €150 per one-way trip, diversion to air would take place in order to save one hour of travel. The value of time is assumed to be €30 per hour, growing in real terms at 1.5 per cent per year. In addition, it is assumed that the airline passenger would incur an additional €8 in access and egress vehicle operating costs to and from the airports. Working through the total private generalised cost in a way similar to previous cases (see Chapter 4, section 1, for example), it would mean that about 12 per cent of the travellers by air would be generated and the rest diverted.

Table 6.5 displays the results of the calculation. This time the calculation assumes that the investment is made at the beginning of the first year, when the aircraft is delivered. To avoid discounting the investment, the calculation includes a year 0. Also, the economic appraisal excludes taxes from prices in order to compute opportunity costs.

---

12   The simplification of not including HSR investment cost is made because an investment in a new railway line is not comparable to an investment in an aircraft. A train normally has a much higher capacity than aircraft serving comparable routes. In addition, the investment in infrastructure on a railway line is route-specific, whereas for aviation, airports are not specific to a given route. Making a direct comparison always involves strong assumptions regarding infrastructure expenditure, including the use of average costs, whereas investment appraisals should be made with marginal costs.

Table 6.5 Returns on investing in an air service

			Year\PV	0	1	2	10	20
				Air				
(1)	RPK	(million)	429.9	0.0	230.0	230.0	230.0	230.0
(2)	Revenues	(EUR m)	429.9	0.0	34.5	34.5	34.5	34.5
(3)	Revs. after tax	(EUR m)	390.8	0.0	31.4	31.4	31.4	31.4
(4)	Op. costs	(EUR m)	312.6	0.0	25.1	25.1	25.1	25.1
(5)	Op. costs before tax	(EUR m)	284.2	0.0	22.8	22.8	22.8	22.8
(6) = (2) − (5)	Gross producer surplus	(EUR m)	145.7	0.0	11.7	11.7	11.7	11.7
(7) = (3) − (4)	Net producer surplus	(EUR m)	78.2	0.0	6.3	6.3	6.3	6.3
(8) through (1)	GHG	(EUR m)	40.1	0.0	3.2	3.2	3.2	3.2
(9) through (1)	Air pollution	(EUR m)	14.3	0.0	1.1	1.1	1.1	1.1
(10) (through (1)	Noise	(EUR m)	28.7	0.0	2.3	2.3	2.3	2.3
(11) = (8) + (9) + (10)	Total external cost	(EUR m)	83.1	0.0	6.7	6.7	6.7	6.7
(12)	Investment	(EUR m)	50.0	50.0				
(13)	Taxes	(EUR m)	4.5	4.5				
(14) = (12) − (13)	Investment net of tax	(EUR m)	45.5	45.5				
				Alternative mode (HSR)				
(15)	RPK	(million)		0.0	205.3	205.3	205.3	205.3
(16)	Revenues	(EUR m)	383.8	0.0	30.8	30.8	30.8	30.8
(17)	Subsidies	(EUR m)	102.3	0.0	8.2	8.2	8.2	8.2

(18)	GHG	(EUR m)	0.0	0.0	0.0	0.0	0.0
(19)	Air pollution	(EUR m)	0.0	0.0	0.0	0.0	0.0
(20) through (15)	Noise	(EUR m)	12.8	0.0	1.0	1.0	1.0
(21) = (18) + (19) + (20)	Total external cost	(EUR m)	12.8	0.0	1.0	1.0	1.0
	Pax consumer surplus						
(22) through (1) × VoT	Diverted traffic	(EUR m)	86.7	0.0	6.2	6.3	7.0	8.2
(23) through (1) × VoT	Generated traffic	(EUR m)	5.2	0.0	0.4	0.4	0.4	0.5
(24)	Access & egress op. cost	(EUR m)	20.5	0.0	1.6	1.6	1.6	1.6
(25) = (22) + (23) − (24)	Net consumer surplus	(EUR m)	71.4	0.0	4.9	5.0	5.8	7.0
	Financial returns						
(26) = (7) − (12)	Airline net cash flows	(EUR m)	28.2	−50.0	6.3	6.3	6.3	6.3
	Airline FRR		11.0%				
(27) = (6) − (14)	Airline gross FRR	(EUR m)	100.2	−45.5	11.7	11.7	11.7	11.7
	Airline gross FRR		25.4%				
(28) = (17) + (27)	Differential flows	(EUR m)	202.5	−45.5	19.9	19.9	19.9	19.9
	Differential FRR		43.7%				
	Economic returns						
(29) = (6) − (14) + (17) + (25)	Without externalities	(EUR m)	273.9	−45.5	24.8	24.9	25.7	26.9
	ERR		54.9%				
(30) = (29) − (11) + (21)	Net economic flows	(EUR m)	**203.6**	−45.5	19.1	19.2	20.1	21.3
	ERR		**42.6%**				

The airline would generate €34.5 million per year (row 2), higher than the €30.8 million of HSR (row 16), due to traffic generation by the airline. Airline revenues and costs are taxed (rows 2 to 5 and 13), unlike HSR, which, in addition, is subsidised (row 17). For HSR, revenues are assumed to cover both operating and infrastructure costs except for the subsidy requirement. The infrastructure costs of the air service are included in the airline ticket price.

The air service produces an operating profit before depreciation (producer surplus) both gross and net of taxes (rows 6 and 7). Comparing this surplus with the investment cost gross of taxes yields the value of the service to the airline (row 26) discounted at the cost of capital of 5 per cent, which is equivalent to a financial return of 11 per cent. The total financial return of the air service would be gross of taxes (row 27), which would constitute the total return to the government should the airline be owned by the government.

Continuing with the assumption that the airline is owned by the government, should the government be able to scale back the HSR service after introducing the air service, the financial return to the government would be higher. This is because the HSR service constitutes a net financial liability to the government, equal to the total annual subsidy. The flows to the government, assuming that each airline passenger is accompanied by a proportional decrease in the subsidy to the railway, are included in row 28, and show that the financial return to the government would be a very high 43.7 per cent.

The economic profitability of the investment includes measures of consumer surplus and externalities resulting from the project. Consumer surplus results from both travel time savings to passengers who divert from HSR to the airline and the willingness to pay for the air services by generated passengers. Without any other economic distortion or any cost economies in secondary markets resulting from the project (see Chapter 2, section 7.3), the consumer surplus incorporates the benefits to the wider economy in terms of productivity gains and, more widely, welfare gains that can be attributed to the air service. The resulting net economic value of the project is €273.9 million (row 29), a large return for an investment of €45.5 million worth of resources, as indicated by an economic IRR of 54.9 per cent.

To arrive at the final return of the project, externalities need to be accounted for. Note that externalities are not just disturbances. They may also register loss of productivity to the bearer. Once externalities are included, the economic value falls to €203.6 million and economic return to 42.6 per cent. The value and returns generated by the project are very large. In NPV terms, the air service generates a net return to society that is over four times the resources invested as capital expenditure. This is value over and above that which would have been generated if passengers had been forced to travel by HSR.

Whereas the objective of the exercise was not to produce empirical results, the orders of magnitude employed were realistic, which means that such high returns perhaps merit some comment. The value of aviation arises from two main factors. Firstly, cost-effective high speed at cruising altitude, which allows substantial time savings beyond a minimum distance of about 500–700 kilometres. Behind the willingness to pay for such time savings lie elements such as personal, commercial and cultural relationships which are better maintained as a result of the investment due to improved access. In addition, generated traffic means that new relationships are established because of the presence of air transport links. The second source of value of aviation arises from the fewer infrastructure requirements of air transport relative to land transport modes, increasing the cost-effectiveness of aviation. Both factors combine to strengthen the advantage of aviation over longer routes, as distance enhances both the speed advantage of aviation and the infrastructure requirements of its land competitors.

Chapter 7
Aeronautics

Introduction

This chapter addresses investment appraisal in a subset of the aeronautics sector, including the manufacturing of civilian aeroplanes and helicopters.[1] It excludes vehicles more frequently associated with aerospace such as rocketry, as well as lighter-than-air craft such as airships. Whereas the underlying economic principles in those sectors are the same as for aeroplanes and helicopters, the former differ in two respects. First, they are more geared to freight transport, where operating cost, rather than travel time, may play a more significant role. Secondly, and more importantly, development programmes and production rates are generally more closely tailored to specific customers than is the case in aeroplane and helicopter manufacturing, requiring alternative considerations in the investment appraisal process.

Aeronautics is a high-tech manufacturing sector, as opposed to the other sectors reviewed in this book, which are all in the service sectors. Traditionally, a distinction in investment appraisal between manufacturing and service sectors involves the presence of inventories of finished products. However, economic appraisal focuses on the long term, looking at the entire life of a project, whereas inventories tend to be a short- to medium-term issue. Also, the fact that aeronautics is a high-tech sector opens the door to greater degrees of uncertainty arising from research, development and innovation (RDI) outcomes,

---

1 The term 'original equipment manufacturer,' or OEM, is not used in the book since the use of the term is somewhat ambiguous. Whereas it seems to have referred originally to manufacturers of components or final products – as its name suggests – it is now also frequently used to refer to final assemblers or value-added resellers. This book uses instead the terms component manufacturer and final assembler.

competitor behaviour and the future operational environment of customers, combined with sunk costs of research and development.[2]

In order to estimate the benefits of a project, it is important to determine who the final market user is. Generally, the appraisal should take as final user the flying passenger or freight customer, unless there is a competitive market that the product would meet first as it trades up the value chain. This consideration is relevant in two respects. First, investments in the aeronautical sector consist of projects involving not only the final assembly and delivery of aircraft but also the building of components, including engines, fuselage parts, avionics, etc. Today most aircraft are developed to be sold competitively in the airline market (itself competitive, in that airlines compete among themselves). Therefore the final user can be taken to be the airlines, rather than the passenger or the freight forwarder. Most aircraft components tend to be tailor-made for specific aircraft. The economic appraisal case should therefore ultimately rest on the aircraft, so that the users are considered to be the airlines.

Where a project consists instead of generic components to be supplied to a multitude of aircraft types or even to other transport modes, the investment appraisal is made independently of the final aircraft. Such investments tend to occur in competitive markets and therefore the ultimate benefit of the project is reflected by the market price of the components (instead of by that of the finished products). The users of the project are the manufacturers that use such components, rather than the airlines, or the passengers.[3]

---

2 Whereas air traffic management (ATM) is also a high-tech sector, it is subject to much weaker competitive pressures than aircraft manufacturing. Similarly, investments in ATM technology, particularly the most innovative elements, tend to be more closely coordinated with technology users, usually involving the public sector.
3 This would be akin to, say, investments in a cement plant to produce cement for the construction industry in general, instead of for a specific construction project. The users in the investment appraisal are the construction companies and project developers, not the dwellers in the buildings they construct.

The second respect in which the identification of the user is important relates to the difference in the treatment of primary and secondary markets. Following from the discussion in Chapter 2, section 7.3, whereas externalities and distortions in the primary market are always included in the economic analysis, distortions in the secondary market are included only when a project affects prices or final quantities in the secondary market. For aircraft assemblers and component manufacturers for specific aircraft, the primary market is the manufacturing of the aircraft that are sold to the airlines. The operation of the aircraft, which is the airline market, constitutes a secondary market (in this case a user market). Other relevant secondary markets include the airport market (a complementary market) and the avionics markets (input markets).

As is the case with airlines, aeronautical projects tend to occur within competitive markets. This is less so for airport projects, where alternative airports may only be available at an additional cost to the user, resulting in market structures that would tend to correspond to models of monopolistic competition. This situation is even more accentuated in the ATM sector. Because of such competitive conditions, the economic analysis of projects in the aeronautical sector is normally based on the financial analysis, correcting for subsidies, taxes or externalities. In cases where the aeronautics project takes place in an imperfectly competitive market, the economic analysis should incorporate the standard adjustments for the effects of increased competition, as will be illustrated in this chapter.

Unlike the airlines, though, entry costs in aeronautics are high, even very high, as is the case in the final assembler market. The launching of a new aircraft model involves heavy investments in RDI. Whereas such investments can be staged, they necessarily build up over long gestation periods into large, sunk investments, coupled with long lead times between the time when orders are placed and when the final product is delivered. In contrast, setting up a new airline can be a question of, say, a year, and the entry into new routes is a process that only takes months. Few investments in the airline industry are sunk (mostly marketing costs and management time), as aircraft are both highly standardised tradable equipment and mobile assets that can be redeployed to alternative markets rather quickly. All this implies that investment in aeronautical projects can be subject to higher uncertainty and risks than investments in airline fleet projects. Such uncertainty, combined with the long, staged development periods involved, mean that real options analysis (ROA) can be a particularly useful tool in the appraisal of investments in the aeronautical sector. Moreover, the economic value of real options may be different from their financial value, with implications for government intervention in the sector.

This chapter is organised around two broad project examples. First a lower uncertainty case, involving the development of a new aircraft model to replace an older aircraft on a tried and tested segment of the aircraft market. The second is the development of a highly innovative component, the market prospects for which are surrounded by a high degree of uncertainty.

1 Low Uncertainty

1.1 Standard Product in (Almost) Perfectly Competitive or Oligopolistic Markets

An aircraft manufacturer is planning to launch a new aircraft model to replace an existing very successful model that is now deemed too old a design and has exhausted upgrade potential. The manufacturer is a well-established brand, with sound after-sales support, trusted by airlines. Indeed airlines have encouraged the manufacturer to launch the new model and a few are willing to help the carrier in the design process, as well as to participate as launch customers. The final assembler can therefore be quite confident that it can generate sufficient orders to make the project at least reasonably successful.

The market is supplied by a small number of aircraft manufacturers, all producing comparable aircraft at a comparable price. The price is set either in an oligopolistic manner, through market signals, custom, price leadership, etc., or competitively.[4] The implication as far as project appraisal is concerned is the same either way, the investing promoter does not have any distinctive pricing power from that of its competitors and will be immediately followed should it deviate from existing prices. The project appraisal therefore takes the sale price as given. In the current case, the aircraft will be sold at €50 million per unit, plus 10 per cent of sales tax.

Total development cost will be €6 billion over six years and it will experience average recurring costs of €20 million per aircraft produced, including test aircraft. The final assembler will be building a production line with capacity for producing 5 aircraft per month or 60 a year, a delivery rate it expects to achieve into the third year of production. It assumes that the aircraft will sell for 15 years before a new version, requiring a new investment programme in product upgrade, is necessary. The analysis ignores residual value after those 15 years because of the uncertainty of the state of the industry in the future and, therefore, of the likely required investments for an upgraded version. The government will assist the development through €300 million of grants for certain qualifying RDI activities.

The calculation and results are displayed under Scenario 1 in Table 7.1. A simple financial analysis shows that the net present value of the project is €7.4 billion (row 8), and that the rate of return for the promoter is 15.4 per cent. The economic return would build upon the financial calculation by making the necessary adjustments for transfers or distortions. In the current case, this consists of adding back to costs the development costs financed through the government subsidy and adding to project benefits the money transferred to the government as taxes.[5]

---

4 For an accessible discussion of market coordination outcomes in conditions of less-than-perfect competition and without communication among players, see Kay 1995. For a formal exposition of alternative models of competition see any textbook on industrial organisation, for example Belleflamme and Peitz 2010 or Martin 2010.
5 Other typical adjustments for this type of project may be removing the cost of taxes on inputs and any shadow price of labour should any of the

The resulting economic net present value (NPV) is €9.2 billion (row 9); and the economic rate of return is 16.5 per cent, slightly higher than the private financial return.

It is worth mentioning in passing that given the combination of low risk assumed to be carried by this investment and the accompanying relatively high financial return would signal an oligopolistic outcome. A perfectly competitive outcome would mean that the financial return would be equal to the opportunity cost of capital, in this case a rate of return of 5 per cent. As far as the economic appraisal is concerned, oligopolistic and perfectly competitive outcomes are treated equally, in that the project does not bring about any price changes in either primary or secondary markets.

The analysis does not incorporate any effect for the externalities caused by the operation of the aircraft. The price paid for the aircraft represents the value of the marginal product of capital (in this case, the aircraft) to the airlines. That is, it reflects the extent to which the output of the manufacturer contributes to the net financial benefit generated by the airline, including both revenues and costs. In the event that externalities are internalised, these will affect both the operating costs of the aircraft and the demand for air travel and such flows will be reflected in the airline's willingness to pay for the aircraft. Where the external costs are not internalised, insofar as the aircraft market is competitive and there are no significant differences in the environmental performance of the different competing aircraft, the airline market is a secondary market where prices and quantities faced by the airlines will not be affected by the project. The amount of pollution with and without the project is therefore the same and has no effect on the economic returns of the project.

R&D or manufacturing activities be located in areas of high unemployment. However, these are general economic appraisal issues (see, for example, de Rus 2010 and Campbell and Brown 2003) not specific to aviation and we ignore them for simplicity. For a broader discussion of shadow prices see Londero 2003.

Still, in cases where externalities are not internalised, it is always worthwhile to test what would be the impact on the project should the airlines be forced to internalise them. This is done in Scenario 2 in Table 7.1. Should government impose taxes on carbon, aircraft noise and air quality, the airlines will pass on these costs to passengers by increasing the price of air tickets, resulting in a decrease in traffic and hence in demand for aircraft. Assuming that the combined environmental taxes would increase average air ticket prices by 20 per cent, and that the price elasticity of demand for air travel is −1.25, the taxes would result in a fall in demand of 25 per cent. This would result in a fall in demand for aircraft, resulting in lower sales (row 10 versus row 1), reduced revenues (row 11) and taxes (row 12), but also lower recurring costs (row 13).[6] The result is that the financial return of the project to the promoter falls from 15.4 per cent to 11.9 per cent and the economic return would fall from 16.5 per cent to 13 per cent. This indicates that, despite no immediate prospect of the externalities in the secondary market being internalised, should they be internalised, the project would still make financial and economic sense.[7]

---

[6] Note that the effect on the primary market (i.e. the aircraft manufacturing market) is a fall in demand (a shift in the demand curve) rather than a fall in the quantity demanded (resulting from an increase in price), since the increase in price occurs in the secondary (i.e. the airline) market. The difference is important because price changes in the primary market have additional welfare implications for the project, as is illustrated in the next section of this chapter.

[7] Note that it is assumed that the introduction of the environmental tax occurs after the aircraft manufacturer has either carried out or at least committed to the installation of sufficient capacity for a delivery rate of 60 aircraft per year. Otherwise, if the manufacturer expected an environmental tax, it would revise downwards its delivery rate and install less assembly capacity. Such a move would help improve the returns of the project somewhat. Still, the returns would be lower than in the scenario of no environmental taxes, since the fall in investment cost would not be proportionate to the fall in production capacity, as project R&D costs would be unaffected.

Table 7.1 Returns on investment in the aeronautical sector under alternative scenarios regarding competition and external costs

		Year\PV	1	2	3	4	5	6	7	8	9	21
	SCENARIO 1: Competitive outcome means project does not alter aircraft numbers on (secondary) airline market											
(1)	Aircraft deliveries	(units)							15	30	60	60
(2)	Cumulative deliveries	(units)							15	45	105	825
(3)	Revenues after tax	(EUR m)	27,636						750	1,500	3,000	3,000
(4)	Taxes	(EUR m)	2,764						75	150	300	300
(5)	Recurring costs	(EUR m)	9,739			50	30	30	310	610	1,210	1,210
(6)	Non-recurring costs	(EUR m)	5,076	1,000	1,000	1,000	1,000	1,000				
(7)	Subsidies	(EUR m)	272	100	100							
(8) = (3) − (5) − (6) + (7)	Net financial flows	(EUR m)	7,405	−900	−900	−1050	−1030	−1030	440	890	1,790	1,790
	FRR		15.4%									
(9) = (8) + (4) − (7)	Net economic flows	(EUR m)	9,195	−1000	−1000	−1050	−1030	−1030	515	1,040	2,090	2,090
	ERR		16.5%									
	SCENARIO 2: Assumes externalities on (secondary) airline market are (unexpectedly) taxed											
(10)	Aircraft deliveries	(units)							11	23	45	45
(11) = ((1) − (10)) × price	Reduced after-tax revenues	(EUR m)	6,898						200	350	750	750
(12) = ((1) − (10)) × tax	Reduction in taxes	(EUR m)	690						20	35	75	75
(13) = ((1) − (10)) × unit cost	Reduction in recurring costs	(EUR m)	2,759						80	140	300	300

(14) = (8) − (11) + (13)	Net financial flows	(EUR m)	4,317	−900	−900	−1050	−1030	320	680	1,340	1,340
	FRR		11.9%								
(15) = (9) − (11) − (12) + (13)	Net economic flows	(EUR m)	5,592	−1000	−1000	−1050	−1030	375	795	1,565	1,565
	ERR		13.0%								

SCENARIO 3: Adjusts Scenario 1 to assume that project increases competition in the (primary) aircraft market Externalities are either ignored or assumed to be internalised

(16) = 10 × ((1) × 19%)	Fall in deadweight loss	(EUR m)	793	0	0	0	0	29	58	115	115
(17) = (9) + (16)	Net economic flows	(EUR m)	9,988	−1000	−1000	−1050	−1030	544	1,098	2,205	2,205
	ERR		17.2%								

SCENARIO 4: Adjusts Scenario 3 to assume that externalities are not internalised

(18) = (2) × 4 × 0.19 × 2	External costs	(EUR m)	4,213	0	0	0	0	23	69	162	1,269
(19) = (17) − (18)	Net economic flows	(EUR m)	5,775	−1000	−1000	−1050	−1030	521	1,028	2,044	936
	ERR		14.0%								

SCENARIO 5: Adjusts Scenario 1 to assume that aircraft is better performing than competitors Environmental costs are not internalised

(20) = (2) × 0.5	Gain in noise efficiency	(EUR m)	1,369	0	0	0	0	8	23	53	413
(21) = (9) + (20)	Net economic flows	(EUR m)	10,565	−1000	−1000	−1,000	−1030	523	1,063	2,143	2,503
	FRR		17.4%								

SCENARIO 6: Adjusts Scenario 4 to assume that the aircraft is better performing than competitors The project increases competition and externalities are not internalised

(22) = (19) + (20)	Net economic flows	(EUR m)	7,144	−1000	−1000	−1050	−1030	528	1,051	2,096	1,349
	ERR		15.2%								

200   *Aviation Investment*

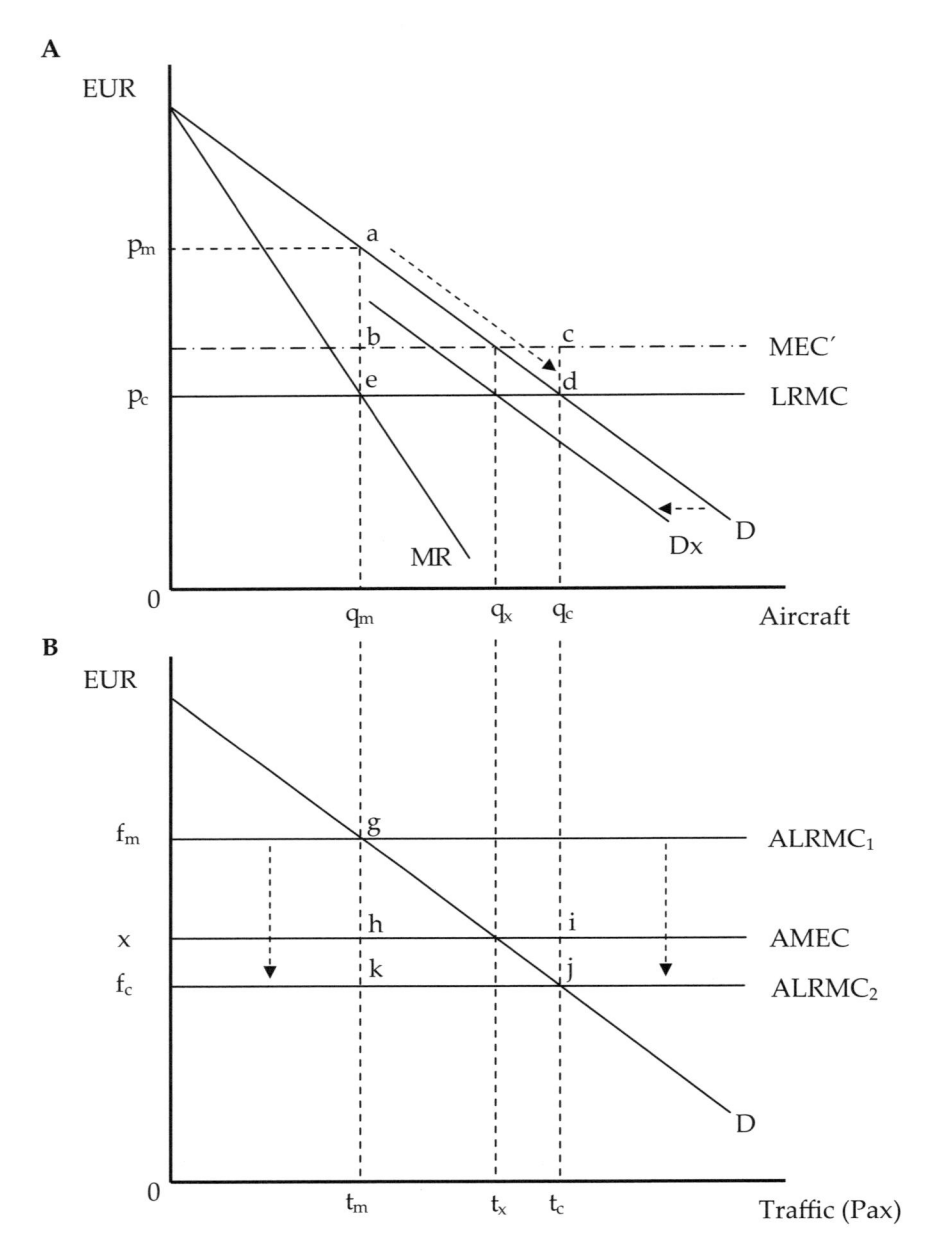

Figure 7.1 Effect of introducing competition in a monopolised aircraft market

1.2 Entering a Monopolistic Competitive Market

The analysis has assumed so far that the project does not alter the total number of aircraft in the market. Should the project not be produced, other well-established aircraft manufacturers would produce aircraft of similar quality at a similar price. An alternative scenario would be that the new aircraft supplied by the project would alter the structure of the aircraft market, affecting the price of aircraft offered to airlines. This could be the case of a project consisting of entering an aircraft segment until then supplied by a monopolist on conditions of monopolistic competition. Normally, in such conditions the monopolist would be enjoying monopoly rents by charging a higher price than the competitive price. The project promoter would intend to bring prices down to a competitive level, forcing the monopolist to follow suit.[8]

The situation is illustrated in Figure 7.1. The upper diagram, 7.1A, corresponds to the aircraft market and the lower diagram, 7.1B, to the (secondary) airline market that makes use of the aircraft. In Figure 7.1A the incumbent monopolist aircraft manufacturer generates super-normal profits by setting the aircraft sale price where marginal revenues (MR) equal long-run marginal cost (LRMC), resulting in the monopoly outcome of price per aircraft p_m and quantity of aircraft supplied q_m. As the project promoter enters the market segment, competition between the two manufacturers brings the price down to the competitive price p_c equal to LRMC, which is consistent with normal profits for the manufacturers, the competitive outcome. The result can be split into two effects. First, there is a transfer

---

[8] Note that a key difference in the underlying assumption between this situation and the oligopoly or competitive outcome reviewed in section 1.1 of this chapter is that in the current case, in the absence of the project the status quo would have remained, whereas in the competitive or oligopolistic case, in the absence of the project competitors would have taken up the capacity otherwise supplied by the project promoter. The market structure and competitive conditions play a crucial role in investment appraisals in sectors where competition is possible, which means that building the 'with project' and 'without project' scenarios must be grounded on industrial organisation models. For further reading on models of competition see Belleflamme and Peitz 2010 or Martin 2010.

of welfare (in the form of income via lower prices) from the manufacturer to the airlines equal to the area $p_m aep_c$. Total welfare to society remains unchanged by this effect. Second, there is a reduction in the deadweight loss that resulted from monopoly pricing. As the lower price of aircraft encourages airlines to place more aircraft orders, there is a net gain in welfare equal to the area ade, plus the increase in (normal) profits in the aeronautical sector resulting from the expanded output.

The impact on the airline market, depicted in Figure 7.1B, is to decrease the airlines' long-run marginal cost (ALRMC) curve from $ALRMC_1$ to $ALRMC_2$. The airline market is competitive and therefore the airlines transfer the welfare gain received from the manufacturers to passengers by lowering fares from f_m to f_c, causing an increase in traffic from t_m to t_c. This results in a final transfer of welfare from the airlines to its passengers of area $f_m gkf_c$ and a net gain in welfare measured by the area gjk, plus the growth in the (normal) profits of the airlines, resulting from the growth in output from t_m to t_c.

The implications for the calculation of project return are illustrated in Scenario 3 in Table 7.1. Scenario 3 assumes that the outcome of the project to the promoter is as described in Scenario 1 – that is, private profitability remains as described in row 8 – and goes on to make the necessary adjustments to the economic returns to incorporate the alternative competitive scenario. Also, it is assumed that airline external costs are internalised or, alternatively, the illustration disregards externalities. The final price of aircraft with the project remains the same as in Scenario 1, at €55 million, but it is assumed that the price charged by the monopolist incumbent before the project was €65 million, a fall of €10 million per aircraft, or a reduction in price by 15 per cent. With a price elasticity of demand for aircraft of −1.25, the fall in prices would cause an increase in the quantity of aircraft demanded, so that about 19 per cent of the aircraft sold by the project would consist of a net increase in sales in the market segment.[9] The reduction in deadweight loss results

---

9 The increase in the number of aircraft with the project is approximate, in two respects. First, for simplicity of presentation the total number of aircraft delivered in Scenario 3 is assumed to be the same as in Scenario 1.

from multiplying the change in quantity demanded (19 per cent of aircraft delivered) from the change in prices in absolute terms (€10 million). There is no need to divide by two, as would be necessary following the rule of a half, as Table 7.1 reflects the changes for the promoter, not the entire market.[10]

The gain by eliminating the deadweight welfare loss is added to the economic benefits of the project. The transfer of welfare from the incumbent (and the government that appropriated 10 per cent of the revenues of the incumbent via a sales tax) to the airlines (area $p_m aep_c$ in Figure 7.1A) and on to the passengers (area $f_m gkf_c$ in Figure 7.1B) is ignored because, as a transfer, it does not change total welfare. The result is then an additional gain in welfare of €793 million (row 16), increasing the net present value of the project to almost €10 billion (row 17) from the €9.2 billion in the scenario where prices do not change (row 9). This would correspond to an increase in the economic returns of the project from 16.5 per cent to 17.2 per cent. The additional welfare gain corresponds to area ade in Figure 7.1A. It ultimately corresponds to area gjk in Figure 7.1B, as the final beneficiaries of the increased competition in the aircraft market are the airline passengers. However, the benefits of the project through reducing deadweight loss are either (preferably) area ade, or (as a surrogate measure) area gjk. Adding up the two areas would constitute double-counting.

In Scenario 3 the project results in a change in the total number of aircraft in the market by virtue of having forced a

---

For ease of computation and reference, the effect of the €10 million price difference on changes in quantity demanded through the price elasticity of demand is calculated as a price increase from the final number of aircraft, rather than as a price fall from the original market size without the project. Second, aircraft cannot be delivered in fractions, so that final demand figures must necessarily be rounded.

10 It is assumed that the former monopolist is left with a 50 per cent share of the market segment. That is, following the entry of the competitor the incumbent does not see its sales fall by 50 per cent, but rather by 50 per cent minus the 19 per cent generation in sales resulting from the fall in prices. Therefore the same reduction in deadweight loss is attributable to the monopolist sales. By not accounting for the welfare gain resulting from the reduction in the deadweight loss attributable to the sales made by the competitor by lowering prices, the calculation effectively incorporates the rule of a half.

fall in the price of aircraft. Clearly, and unlike in Scenario 1, where the total number of aircraft did not change with the project, the question of whether externalities in the (secondary) airline market are internalised or not becomes important for the estimate of economic profitability, since any change in external costs in the secondary market as a result of the greater number of flights enabled by the project, can be attributed to the project. Scenario 4 builds upon Scenario 3, but assumes that externalities in the airline market are not internalised. That is, Scenario 4 assumes that the outcome of the project as far as the promoter is concerned is as described in Scenario 1, implying that the net present value of private financial flows to the promoter is as described in row 8, but makes the necessary adjustments to the calculation of economic profitability to incorporate the new assumption about externalities. The airline marginal environmental cost (AMEC) caused by the aircraft is depicted by the AMEC schedule in Figure 7.1B. As seen when discussing Scenario 3 above, the project has brought about an increase in the total demand for aircraft from airlines from q_m to q_c, in tandem with an increase in traffic from t_m to t_c and a reduction in airline fares from f_m to f_c. However, the extra passengers (t_c-t_m) are not paying the full cost of their air travel, as they are causing an environmental externality ($x-f_c$) per trip, resulting in a total external cost equal to area hijk. This is a direct consequence of the project and therefore needs to be taken into account when measuring economic returns.

Alternatively, the external environmental cost of the project could be measured through the aircraft market as area bcde in Figure 7.1A. MEC stands for marginal environmental cost per aircraft, and the MEC line is dotted and an apostrophe added to underscore that this is an alternative, parallel approach, basing the calculation on emissions per aircraft rather than emissions per passenger. Adding areas bcde in Figure 7.1A and hijk in Figure 7.1B as costs in the analysis would result in double-counting the environmental costs of the project.

The total annual environmental cost of operating an aircraft is estimated at €4 million per year, including emissions of greenhouse gases (GHG), air pollutants and noise, all of which are assumed not to be internalised, remaining as an external cost. Area hijk in Figure 7.1B would correspond to the €4 million cost per aircraft, multiplied by the total number of aircraft in operation in the market segment that can be attributed solely to the project and that would not have existed had the project not taken place. This latter figure would be 19 per cent of the cumulative deliveries, where 19 per cent is the estimated traffic generated by the project. The figure must account for all aircraft in the segment, meaning that it must include those of the other competitor, which is assumed to retain a 50 per cent market share. The external environmental cost each year is therefore twice 19 per cent of the total cumulative aircraft deliveries, times €4 million (row 18). Note that, as demand for air transport grows, the demand curves in Figure 7.1A and B would shift rightwards every year, implying that area hijk in Figure 7.1B would grow every year.

The estimate of economic profitability of the project (row 19) now combines the gain in avoided deadweight loss estimated in row 16, brought about by the increase in aircraft in operation in the market segment as a direct result of the project, with the environmental cost that such additional aircraft bring about in the secondary market (row 18). The economic value of the project falls considerably, by €4.2 billion, to €5.8 billion (row 19), and the economic return from 17.2 per cent to 14 per cent. Note that the net financial value of the project to the promoter would still be €7.4 billion (row 8) and the financial rate of return 14.4 per cent, both higher than the economic return. The difference is explained by the environmental externalities, the business gained from the incumbent and the government subsidies.

If, once the investment had been carried out, the government suddenly introduced taxes internalising all of the external costs of aviation, the adjustment to the estimation process would be akin to that in Scenario 2. The taxes would mean that airline fares would rise from f_c to x in Figure 7.1B, and the LRMC schedule of airlines would effectively become the AMEC schedule. The resulting quantity demanded of air travel would be t_x. The impact on the aeronautics market of such an increase in airline costs and fares would be that the aeronautics industry would face a fall in demand, represented by the move from schedules D to Dx in Figure 7.1A. For an aircraft price of p_c, the quantity demanded of aircraft would be q_x. The estimates of financial and economic return would have to be adjusted to incorporate the lower number of aircraft deliveries and the external cost would disappear as a separate item from the estimate of economic profitability.

1.3 Entering a Market with an Improved Product

Finally, it is worth considering the case of producing an aircraft that yields improvements in operating performance relative to older models. Let us assume that the performance is environmental in terms of emissions of GHG, air particles or noise. Following the discussion above, if the regulatory framework of the airline industry is such that external costs are internalised, when calculating the economic returns there is no need to make any consideration different from those made for estimating the financial return. The improved environmental and economic performance will be reflected in the demand for the aircraft. If all competing manufacturers produce aircraft with the same improved performance, the economic return will be equal to the financial return and will be equal for investments across all manufacturers. This would correspond to the lower left quadrant in Figure 7.2, which sets out the relationship between financial and economic returns in competitive markets, as determined by whether environmental costs are internalised or not, and by whether the promoter differentiates its output from the competition through a distinctive product.

It may well be that the project manufacturer has a distinctive capability that enables its product to be uniquely high-performing environmentally relative to those of its competitors. Continuing with the scenario that all external costs are internalised, the financial and economic returns of the project would continue to be equal. But the returns would be higher than those of competitors, as the competitive advantage would be reflected by means of a higher market share or the aircraft commanding a higher price. This situation corresponds to the upper left quadrant in Figure 7.2. The fact that the product is based on a distinctive capability means that other competitors will not be able to replicate the performance for the foreseeable future. In competitive markets such a situation is not sustainable over the long run. If a manufacturer is permanently superior it will end up capturing the entire market. Competitors will tend to exploit their own distinctive capabilities to provide value to the airlines, equating investing performance over the long run, or being driven out of the market.

The outcomes would change if externalities were not internalised in the airline industry. Then there may be differences between the financial and the economic return. Again, let us consider two possible scenarios regarding differentiation in product performance relative to those of competing manufacturers. Firstly, if other aircraft manufacturers produce models with the same improvements in performance, no improvements in environmental performance can be attributed to the project. In the absence of the project, the environmental performance of the airline industry would be the same as with the project, due to the improved aircraft produced by the competition. In terms of the mechanics of estimation, as far as the (secondary) airline market is concerned, the 'without project' scenario enjoys the same environmental performance as the 'with project' scenario. As far as environmental factors are concerned, the financial and economic return of the project would be the same. This would correspond to the lower right quadrant in Figure 7.2. The implication of this conclusion is that in an investment appraisal exercise under competitive conditions and where externalities are present in a secondary market, regardless of how much better the environmental performance of the product is relative to preceding generations, no environmental benefit should be assigned to the investment, since the product does not make a difference to the secondary market. The two key assumptions here are that the primary market is competitive and that the output of different participants in the primary market are perfect substitutes, that is, the participants offer the same performance at the same price.

	Internalised environmental costs	External environmental costs
Distinctive product	FRR = ERR ... and different from those of competitors over the short run.	**FRR ≠ ERR** ... and sustainable over the long run.
Standard product	FRR = ERR ... and equal across competitors.	FRR = ERR ... since the amount of pollution in the secondary (airline) market is unaffected by the project.

Figure 7.2 Treatment of environmental costs in the economic appraisal of investments in competitive aircraft manufacturing markets

The second possibility while externalities are not internalised is that the project differentiates itself from competitors by having a better environmental performance. This would correspond to the upper right quadrant in Figure 7.2. As the improved performance concerns an environmental cost that is not internalised, airlines may not be willing to pay for it and competitors may not even seek to match the environmental performance of the product. Other things being equal, the aircraft with the better environmental performance would only command the market price and the same market share as any other competing aircraft. In such situations, the financial return will be the same for all competing aircraft manufacturers. However, the economic appraisal of the project would require special consideration, especially since, unlike the situation in the upper left quadrant, the current situation is sustainable over the long run.

Scenario 5 in Table 7.1 illustrates the situation of the upper right quadrant in Figure 7.2. Assume that the environmental performance concerned is noise and that the aircraft is slightly quieter than those of competitors, reducing external costs by an estimated €0.5 million per year per aircraft. In a competitive market (either perfectly competitive or an oligopoly) such external benefits would accrue to all aircraft sold by the promoter. The welfare gain resulting in reducing the noise externality amounts to €1.4 billion (row 20). This constitutes an improvement in the economic value of the project that would now total €10.6 billion (row 21), up from the €9.2 billion of Scenario 1 (row 9). The financial return to the promoter remains as in Scenario 1.

Likewise, in the context of an aircraft that ends the monopoly of the incumbent manufacturer in a particular market segment (that is, a market characterised by monopolistic competition before the project), the benefit would also apply to all aircraft sold by the entrant manufacturer. In terms of Figure 7.1 the introduction of the new aircraft would constitute a downward shift in the AMEC curve, producing an area equal to €0.5 million along the vertical axis times t_c (or q_c) on the horizontal axis. Since the project produces a distinctive product that would not be replicated in the 'without project' scenario, the gain in environmental performance from the project applies to all aircraft sold by the project, whether substitutes from the incumbent, or generated through lower prices (q_c-q_m in Figure 7.1A). The situation is illustrated in Scenario 6 in Table 7.1. The result is that the economic value of the project improves relative to that of Scenario 4, bringing net economic value to €7.1 billion (row 22) and improving the returns of the project from 14 per cent to 15.2 per cent. The financial value of the project to the promoter remains unchanged and as estimated in row 8.

2 High Uncertainty

2.1 *An Innovative Project Contingent on External Developments*

A manufacturer of aircraft engines is considering investing in the development of a new engine that will yield a substantial performance leap in terms of life-cycle costs in general and fuel consumption in particular. The commercial prospects, however, depend on two critical factors. First, the new engines would require significant changes in the way aircraft are designed, involving heavy investment on the side of final aircraft assemblers, which cannot be taken for granted. This would in turn be influenced by the second factor, which is future government regulation on aircraft operating and emissions standards. The outlook regarding these two interrelated factors is highly uncertain, but management expect these issues to be resolved within the next five years.

Developing the engine will require substantial research and development (R&D) investments over a prolonged period of time including, in present value terms, €1.5 billion in research and an extra €500 million to adapt existing manufacturing facilities, bringing total project investment to a present value of €2 billion. Should the project fail, or be abandoned, practically all of the R&D investment will be lost with no obvious alternative use for the promoter. The promoter can only be sure of selling some of the facilities for a present value of about €15 million. The investment is therefore to be treated as a non-recoverable or sunk cost. An adverse scenario whereby the promoter would spend €2 billion and receive only €15 million in return could potentially bring the engine manufacturer to bankruptcy and would certainly prompt a replacement of top management. On the other hand, should developments in the regulatory and airframe assembler sides evolve favourably and no competitor develop similar engines, the competitive advantage of the manufacturer would be virtually impossible to match for a number of years, until competitors developed comparable engines. This would give the promoter a first-mover advantage that could prove extraordinarily lucrative. The project could easily generate free cash flow with a present value of €10 billion.

The management of the engine manufacturer carried out a discounted cash flow (DCF) analysis of the investment, including a very wide range of scenarios, reflecting the uncertainty surrounding the project. There was a wide variety of opinion and the median scenario was finally chosen, including a relatively modest sales projection by assuming that competitors may develop engines that could potentially tame the prospects of the project. Because of the high risk, the cash flows were discounted at 25 per cent, including a substantial risk premium, compared to the firm's standard target return on investment of 15 per cent and the risk-free rate of 5 per cent. The result of that median scenario was a present value of free cash flow before investment of €1.2 billion. As the present value of launching the project now is €2 billion, the project would have a financial net present value of a negative €800 million, so that management was bound to reject the project.

However, those in the management team who feel most strongly for the project objected that, given the substantial uncertainty and the wide range of possible outcomes, relying on the median forecast alone offers too narrow a guide to the range of opportunities that lie ahead.[11] They argued that whereas standard DCF analysis shows that the project is not worth undertaking as things stand, it may still be worthwhile to keep open the option of launching the project and delay taking the decision until there is less uncertainty about future prospects. They therefore proposed to complement the DCF analysis with an estimate of the option value of the project at the current moment. The objective would be to find out whether it is worthwhile to spend money today to start up the initial phases of the project – thereby keeping the option open – and, if so, how much.

Management decided to estimate the option value in a way that made more visible their ability to exercise it before the expected expiry date in five years' time, opting for the binomial method.[12] Whereas this method allows them to model changes in input variables such as volatility and investment costs

---

11 In addition they may argue that by not doing the project, they run the risk of another competitor taking a lead with the same or similar technology, curtailing the promoter's future position to one of follower. The boundaries of the investment question can be widened, therefore, to simulate potential outcomes for the firm as a whole. The line of reasoning would be similar but the scenarios modelled differently.

12 The binomial method is an approximation to the Black–Scholes formula used in Chapter 6, section 2.2, on airline fleet expansion. The binomial method is more transparent, more flexible about the construction of scenarios and better suited to American options (those that can be exercised at any time before expiry). The Black–Scholes formula addresses European options that can only be exercised at a pre-specified date. See Kodukula and Papudesu 2006.

(the latter being the strike price of the option) throughout the life of the option, they decided to carry out the simplest possible estimation.[13]

The first step consists of estimating of the implied volatility resulting from the scenarios put forward by management which, following the same method illustrated in Chapter 6, section 2.2, regarding options on aircraft, would be estimated as follows:

$$\sigma = \frac{\ln\left(\frac{S_{opt}}{S_{pes}}\right)}{4\sqrt{t}} = \frac{\ln\left(\frac{10{,}000}{15}\right)}{4\sqrt{20}} = 36.35 \, percent$$

... where σ is the volatility of the cash flows of the project, that is, the sale of engines (or the full service programme if applicable), S_{opt} is the underlying asset value under the optimistic scenario, or €10 billion as mentioned above, S_{pes} is the underlying asset value under the pessimistic scenario (€15 million), and t is the period over which volatility is estimated, or project lifetime (20 years). The resulting volatility of 36.35 per cent underlines the high degree of uncertainty surrounding the project.

With the estimate of volatility, the up and down factors and the risk-neutral probability can be estimated, which will enable the building of the binomial tree. The calculation of the three items proceeds in turn, as follows:

$$u = e^{\sigma\sqrt{\delta t}} = e^{0.3635\sqrt{1}} = 1.4383$$

---

[13] Modelling the precise conditions of the options embedded in a project can potentially become a computationally burdensome exercise. As in any other project appraisal exercise, it is up to the analyst to decide how much detail is worth going into and whether relatively simplified estimates can give useful insights. The objective here is to illustrate the use of the real option analysis method. For more complex modelling the reader should consult the specialist literature. For modelling investments under imperfect competition, a topic particularly relevant to the aeronautical sector, see Smit and Trigeorgis 2004.

… where u is the up factor and δt the time associated to each step in the tree, in this case 1 year;

$$d = \frac{1}{u} = \frac{1}{1.4383} = 0.6952$$

… where d is the down factor; and:

$$p = \frac{e^{r\delta t} - d}{u - d} = \frac{e^{0.05 \times 1} - 0.6952}{1.4383 - 0.6952} = 0.4791$$

… where p is the risk-neutral probability and r the risk-free interest rate, in this case 5 per cent

Figure 7.3 presents the binomial tree or lattice for the project. Each column represents a year, starting from column 0 at the left, representing the present, and ending with column 5 at the right, representing the maximum life of the option, namely 5 years. Each cell within each column represents a possible outcome. The calculation of the binomial lattice begins with the estimated present value of future cash flows (before investment cost) that management takes as its central case scenario, namely €1.2 billion. This is the current asset value S_o, in the far left cell, in column 0. Subsequent cells are named with up (u) and down (d) identifiers, with the exponential representing the cumulative number of moves up or down followed to arrive at that cell.

Starting from cell S_o the asset values in successive years are estimated using the up and down factors (u and d, respectively) estimated above. So for the first year, the up and down asset values in thousands of euros are:

$$S_o u = 1{,}200 \times 1.4383 = 1{,}726$$

$$S_o d = 1{,}200 \times 0.6952 = 834$$

The same method is followed to estimate the asset values for successive cells of the binomial tree.

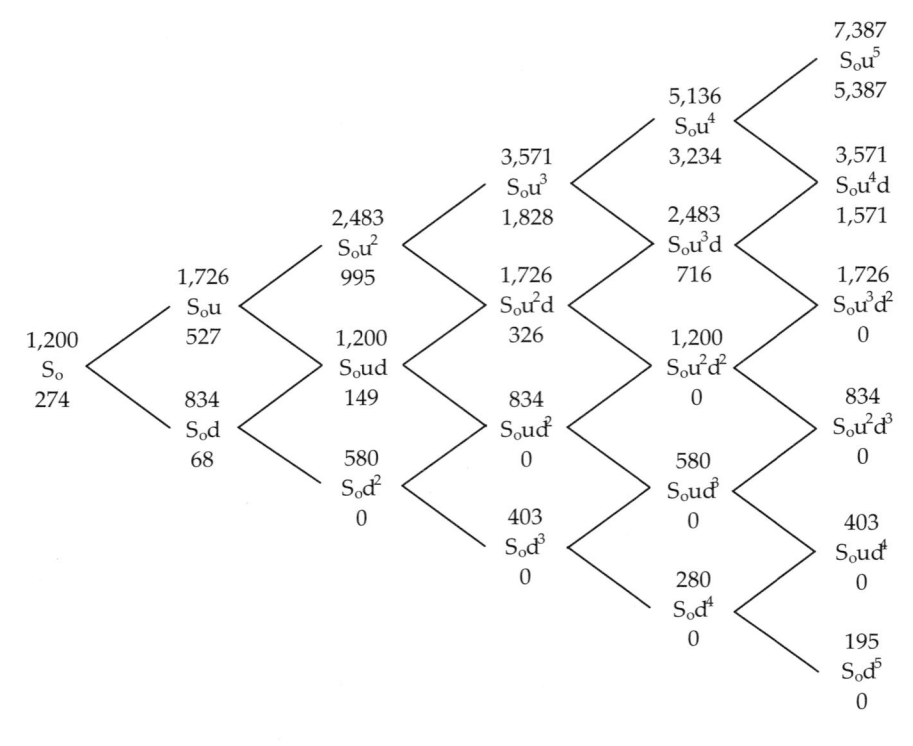

Figure 7.3   Binomial tree for the financial real option value of an aircraft engine project

The real option value (ROV) for each cell is calculated once the asset values (the upper figure in each cell) are calculated for all cells, and is included at the bottom of each cell. The calculation of option values starts from the right end of the binomial tree, that is, the last column – column 5 in the current example – following a process known as backward induction. The calculation procedure for the final column differs slightly from that for all the other columns preceding it.

At the final column – including what are known as the terminal nodes or terminal cells – the uncertainty surrounding the future revenues of the project is expected to be resolved. By then, the government will have reached a decision regarding the regulation of emissions standards and airframe manufacturers could therefore be expected to have reached a decision as to whether to develop new aircraft that would accommodate the new engine technology being contemplated by the project promoter. At that point the decision to make the investment in the new engine must be taken. Given the way the project has been defined, whereby no further delay is possible, waiting has no value. The value of the option is simply the value of the project (after investing).[14] The investment will be made if the value of the expected cash flows exceeds the €2 billion investment cost (where the latter is also the strike price of the option). At $S_o u^5$, for example, the cash flow is worth €7.4 billion which, after an investment of €2 billion, would render a net present value for the project of €5.4 billion, which is the value of having the option to invest at that moment. In cell $S_o u d^4$ the present value of cash flows is €403 million. An investment of €2 billion would have a negative NPV of €1.6 billion, therefore the investment will not be made and the option to invest at that moment is worth zero.

In effect, the value of the option at the bottom of each cell represents the value-maximising decision at that stage. The possible decisions include (i) investing, (ii) keeping the option alive, or (iii) abandoning both the option and the project. Each of these possible decisions is discussed in turn. First, if the value of carrying out the investment at that point (that is, the NPV, or the asset value minus the investment cost) is higher than the value of the option (the value of waiting) at that point, the investment will be made and therefore the value of waiting is 0. The option value becomes the value of the project and therefore the lower figure in the cell includes the NPV of the project. Note that making this decision of investing in the project or not is inescapable in column 5, the final column, since by then the option expires – it cannot be kept alive any longer. In that column the lower figure in each cell either includes the NPV of the project if it is carried out, which is the value of the option at that point in time conditional on being exercised then, or 0, which is the value of the option if the project NPV is negative and hence not worth carrying out.

---

14 That is, if the option to invest could be traded, it would be traded at the NPV of the project. It is useful to think of the value of the option at this stage as akin to the value of a notional licence to carry out the project.

Second, if the value of the option is higher than the NPV of the investment, it will be rational to wait, that is, to keep the option alive, and the cell includes the option value. For example, in cell $S_o u^3$, at the top of column 3, the present value of the project cash flows is €3.6 billion which, after the €2 billion investment cost would result in a project NPV of €1.6 billion. The option value is higher, at €1.8 billion, or, in other words, it is better to wait, therefore the option is not exercised and is kept alive. The cell incorporates the option value, rather than either the NPV of the project or 0. Finally, if the NPV of the investment is negative in that cell and the option is worthless, the option value inserted in the cell is 0. It is worth abandoning, or killing, the option (unless it is free of charge) as well as the project.

As has been seen above, in the last column – column 5, when the option expires – the value of the real option is either the project NPV or 0. For all preceding columns – columns 0 to 4 –, the value of the real option at each cell corresponds to the weighted average of all possible future real option values, estimated from the real option values in the subsequent up and down cells. As an example, the option value at cell $S_o u^4$, at the top of column 4, would be as follows:

$$ROV_{S_o u^4} = \left[p \left(ROV_{S_o u^5} \right) + (1-p) \left(ROV_{S_o u^4 d} \right) \right] \cdot e^{-r\delta t}$$

… resulting in the following value:

$$ROV_{S_o u^4} = \left[0.4791(5{,}387) + (1 - 0.4791)(1{,}571) \right] \cdot e^{-0.05 \cdot 1} = 3{,}234$$

The binomial tree is completed backwards (leftwards) by applying the above formulas to each preceding cell. Column 0, in the far left of the binomial tree, includes the current value of the real option, which is €274 million.[15] The value of the option

---

15 The value of the real option calculated with the Black–Scholes method is €266 million. It is normal for such small differences to occur, as the binomial method is an approximation to the Black–Scholes result, while adding more transparency and flexibility in the definition of project scenarios.

is positive and higher than the NPV of the project at this point (which is negative: €1.2 billion − €2 billion = −€0.8 billion), therefore it is worth keeping the option alive.

Using the traditional DCF analysis, the engine manufacturer would not carry out the project as the NPV is negative. However, the uncertainty embedded in the DCF analysis masks a wide array of possible results, including very profitable outcomes dependent on events that will happen in the future and which could make the project very profitable. Based on the analysis of future possible returns, at this stage, it is worthwhile to pay up to €274 million to keep open the possibility of carrying out the project within the next five years. The investments carried out in keeping the option alive could consist of starting the early development phases, including hiring specialist personnel, developing initial design concepts, etc. Such investments to keep the option alive would be developed further, or abandoned, depending on how events evolve as time progresses. The option value at each point in time indicates the maximum amount that it is worth spending in order to keep the option alive.

Three observations can be made at this point. First, looking at the binomial lattice, it seems rather unlikely that the project will be carried out. By year 5, only in two out of six future scenarios is the project worth undertaking. Those who object to the project may use such results to claim that it may be better not to waste money in keeping the option alive. Still, the rational thing to do is to keep the option alive. The option gains its value from the potentially very large returns should circumstances over the next five years evolve in a way that would favour the project.

Second, even if events develop in such a way that a DCF analysis makes the project viable at some point in the future, it may still be worth waiting rather than proceeding with the project. This is the case depicted in cell $S_o u^2$ in column 2, for example. The estimated asset value in that situation would be €2.48 billion. At the strike price of €2 billion, this means that the project is expected to have a positive NPV of €483 million. Still, the option is worth €995 million. This signals that, whereas the DCF analysis signals that, on a risk-weighted basis, the project is already worth investing in, the degree of uncertainty about the future is such that being able to wait until the future reveals more about the likely outcome of the project is worth more than the expected NPV of the project. Therefore it is still worth waiting rather than investing. In fact, the binomial tree shows that it will be worth waiting to make the investment decision until the option expires in year 5.

Finally, the third observation consists of an extreme version of the preceding observation, depicted in cell $S_o u^4$ in year 4 in Figure 7.3. The project managers may conclude that since the option value (€3.2 billion) – or in other words, the amount that would be worth spending to keep the option alive – is higher than the total investment cost (€2 billion), it may be worth carrying out the project at that point anyway. However, the value of carrying out the project at that stage (€5.1 billion – €2 billion = €3.1 billion) is still less than the option value. This reveals that, if waiting involves a sufficiently low opportunity cost, it is better to wait. After all, one thing is how much one should be willing to pay and another thing is when to pay it. If little is gained by bringing the decision forward, one may as well wait and make the decision under a greater degree of certainty.

2.2 Financial versus Economic Real Option Value

The analysis in the preceding section focused exclusively on returns to the aircraft engines manufacturer. It did not address socio-economic value. Let us suppose that the government wishes to carry out an economic appraisal of the proposed investment project, which they plan to do by building upon the financial appraisal performed by the private promoter. The private financial analysis carried out by the manufacturer of aircraft engines would require three adjustments.

First, the economic analysis would need to add back sales taxes on inputs and outputs paid by the promoter to project benefits. Let us assume that taxes are such that the value of the project before investments would increase from €1.2 billion, to €1.4 billion. The effect of increasing the pay-off would be to increase the value of the real option. At the same time, the capital investment cost would decrease when converted to economic terms by deducting taxes. Let us say that the economic cost of the capital investment would be €1.9 billion instead of the €2 billion in the financial evaluation.

Second, like most RDI investments, the project is likely to generate spillover effects through knowledge creation which could have applications either in aeronautics or in other sectors. By definition such benefits are a (positive) externality and not taken into account in the financial return calculations of the promoter, as any internal benefit would be. Assume that such benefits would amount to an extra €100 million, bringing the present value of the project before investment from €1.4 billion to €1.5 billion. That spillover knowledge would be available whether the project succeeds or not, which would mean that the worst-case scenario would consist of a higher benefit. This could be used to argue for a lower volatility of returns, depressing the real option value. However, it could equally be argued in turn that should the project succeed, the positive payoff could also be higher, keeping volatility constant. The answer to this issue is obviously project-specific. In the current case, for simplicity it is assumed that volatility stays constant.

Third, there are several reasons why the government may use a lower rate of discount to evaluate investments than the private sector.[16] One reason is that if the project is sufficiently small relative to the size of the economy, the government (and society) would have a greater ability than the private sector to bear the non-diversifiable risks inherent in the project, as the risk would be small relative to the size of the economy. Another reason is that capital markets may be subject to distortions such as taxes, which may discriminate between the public and private sectors. Also, the product market where the project takes place may be imperfectly competitive and individual firms

---

16 Social discount rates and their relationship to private or market discount rates are standard topics in any book on cost benefit analysis. Accessible discussions on this topic are included in Boardman et al 2014, de Rus 2010 and Campbell and Brown 2003. For a discussion on a developing country context see Brent 1998.

may demand higher rates of return than would be the case in more competitive markets. Another is that the government may wish to address inter-generational externalities or other ethical considerations by lowering the discount rate for benefits and costs in the distant future. Whichever discount rate is applied by the government, particularly when acting on a tight budget constraint, a good reason should be given for it to be deemed lower than the long-term real rate of interest of public debt, as this rate reflects the marginal cost of funds to the government and society. In practice, social discount rates applied by the government or, if estimates are not available, long-term government borrowing rates, tend to be lower than discount rates applied by private firms. The effect of using a lower discount rate to value the stream of future flows of benefits (before investment) is to increase their value.

Let us assume that in the current project, the lower discount rate applied by the government would result in the value of the benefit flows increasing from €1.5 billion to €2 billion. The value of the project would be €2 billion and of the investment, as mentioned above, €1.9 billion. The project would have an economic net present value of €100 million which, by virtue of being a positive economic value combined with a negative financial value (the €800 million loss identified in section 2.1 above), would render the project as a candidate for government support. Whether it would actually merit financing would depend on the government budget constraint and the socio-economic profitability of alternative projects. Such a low economic return (€100 million for an investment of €1.9 billion) would likely make it a borderline project. However, the government recognises, just as the private promoter did, that there is a large degree of uncertainty surrounding the benefit stream. The returns of the project may be much larger than the mean or expected return and therefore a real option analysis may reveal more value in the project.

Figure 7.4 includes the binomial tree calculated using the economic flows rather than the private financial flows, including the three adjustments mentioned above. To recap, the adjustments had the result of increasing the present value of the asset to €2 billion from the €1.2 billion of the private sector financial analysis; and of decreasing the total investment cost (the strike price of the option) from the €2 billion borne by the private sector in the financial analysis to an economic investment cost of €1.9 billion. The result shows that the economic real option value of the project, at €865 million, is substantially higher than the private financial option value of €274 million (Figure 7.3). If for, say, budgeting reasons or indivisibilities in required investment effort, the €274 million private real option value were not enough for the private sector to keep the option of carrying out the project alive, there may be a strong case for the government to help finance the option.

Moreover, there may be an economic case for the government to help finance real options to the private sector even when the option value to the private sector is 0. This is signalled by the circled lower figures in cells $S_o d^2$, $S_o ud^2$, and $S_o u^2 d^2$ in Figure 7.4, which display positive economic real option values on the project, in situations where the private real option is worthless (same cells in Figure 7.3).[17] Such a result would help justify the case for the government to support the financing of research programmes with commercial prospects too uncertain to be of any value to the private sector, but which may generate such a value in the future depending on the development of events.

In the current example, real option analysis strengthens the case for a project with an economic NPV that was positive but borderline. It is worth pointing out that the situation described in section 2.1 of this chapter in the context of private financial value, where real option analysis was applied to a project with a negative NPV, would also apply in the context of socio-economic value. That is, there can be valid cases for public-funding research programmes that keep options alive on projects for which standard socio-economic appraisal finds the prospects too uncertain to justify carrying them out at the time of the appraisal, but which have positive economic real option values.

---

17 Note that the lower figure in cell $S_o u^3 d^2$ in the fifth column of Figure 7.4 is not circled because it consists of an expected project net present value, or the value of the option at expiration, where delaying the project further is not possible. See section 2.1 of this chapter.

This would be the situation depicted in Figure 7.3, assuming that the values used for the private sector example represented economic values.

Finally, note that the case described has implicitly assumed that the governmental body in charge of reaching the decision on whether to support the option is independent of the governmental body deciding on the regulation of future emissions standards. In effect, it has been assumed that the private sector and the government both face the same degree of uncertainty. This may not be so in practice. In fact, an alternative for the government to providing research grants to help keep options open would be to provide more regulatory certainty.

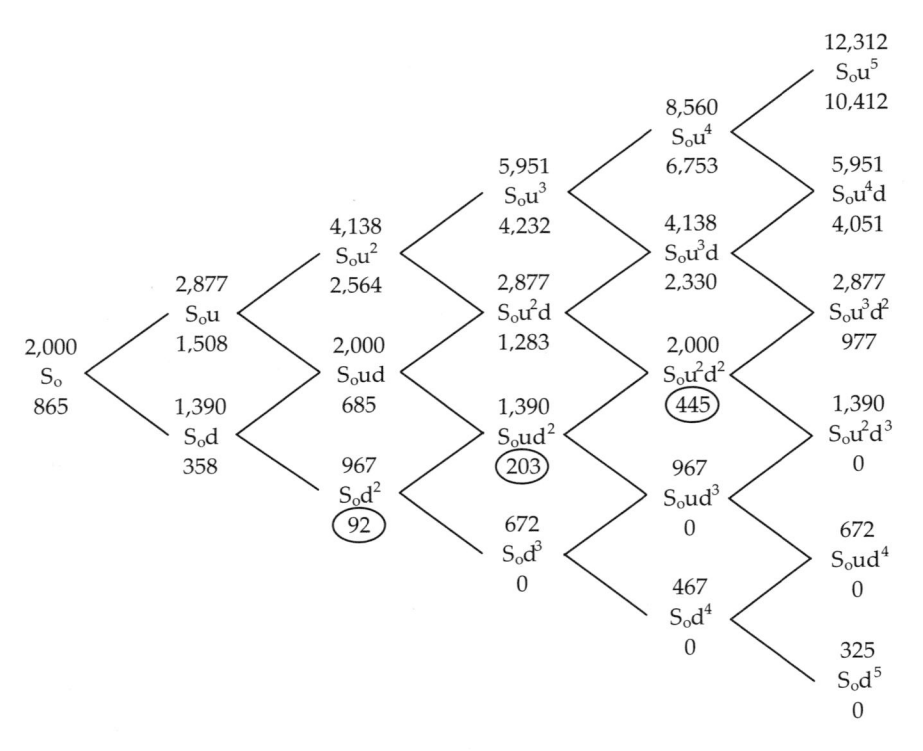

Figure 7.4 Binomial tree for the economic real option value of an aircraft engine project

Chapter 8
Concluding Remarks

The presentations of appraisal methods in this book have focused on identifying costs and benefits, measuring them, and avoiding double-counting or neglect. Inevitably, all appraisals are based on models, and models are simplifications of reality. Models can always be made more detailed in an attempt to reflect a more accurate representation of actual conditions. In addition to the simplifications listed in Chapter 1, section 4, four possible dimensions along which to add detail to the models include benefits, costs, timing and strategic interaction.

Regarding project benefits, perhaps the most fruitful area for refinement concerns delay to users, including both measuring the actual delay caused to users and the cost of such delay. In the case of airports, more accurate delay functions can be constructed with data and simulations performed in the process of project planning and facility design. Relevant items that may be addressed include facility utilisation and capacity constraints in the terminal, user access and egress travel profiles, travel conditions on alternative airports and modes of travel, and airline behaviour in the presence of capacity constraints. Also, more research is needed to better understand the impact of congestion and the ensuing delays both on user behavior and on delay propagation through the air transport network.[1]

Models of airline behaviour become particularly relevant for air traffic management (ATM) projects. These go hand in hand with estimations of airspace capacity and likely delay profiles. Generally, the data to perform such simulations are likely to be available only from the air navigation service provider (ANSP). Some ANSPs and ATM multilateral agencies have

---

1 See, for example, Federal Aviation Administration 2010, regarding delay propagation.

well developed simulation tools for planning and appraisal purposes.[2]

For airline appraisals, established airlines generally have databases with substantial evidence on passenger behaviour across various fare categories. These can be used for estimating the traffic effects of network changes that may result from the introduction of new aircraft.

Reliable estimates of user willingness to pay to reduce trip duration are important for aircraft manufacturers in estimating the underlying demand potential for new products, particularly when they are innovative. The money value of time is central to inform decisions about: (i) whether to produce smaller aircraft aimed at direct services between secondary airports, or larger aircraft serving hub networks; (ii) whether to go for faster, more comfortable but more expensive regional jets versus turboprops; (iii) the extent to which engine technology should prioritise fuel-saving over speed; or (iv), and more innovatively, whether to invest in more expensive aircraft that fly closer to, or beyond, the sound barrier. The analyst may wish to enhance estimates of values of time readily available from governmental agencies with further analysis on variance within the estimates and on how values of time may change with income levels. For example, the analysis of variance in values of time would be helpful when justifying investments in private aviation.

Regarding the cost estimates in the appraisal models, the underlying conditions assumed should reflect the applicable cost economies, on which there is plenty of evidence in the academic literature. Aviation, like any transport infrastructure or vehicle operation, enjoys economies of scale (lower unit costs through larger capacity), density (lower unit costs by using existing capacity more intensively), and scope (lower unit costs by sharing existing capacity to produce different products). Such economies will affect the unit costs resulting from projects that change physical capacity, and the resulting impact on costs

---

2 See, for example, European Model for Strategic ATM Investment Analysis (EMOSIA) by Eurocontrol: http://www.eurocontrol.int/ecosoc/public/standard_page/emosia.html (accessed: 27 July 2013).

may at times be important in determining project viability. The failure to recognise scope and density economies tends to lie behind the often flawed – yet frequent – proposals for tourist-dedicated airports, freight-dedicated airports, or dedicated business-class airlines. Similarly, any scale economies resulting from larger facilities should be set against the time cost to users caused by the accompanying longer throughput processing time. Such an exercise would require sound estimates of facility operating costs, passenger processing time, and the value of time for affected passengers.

Project timing and phasing are important drivers of investment performance and, more generally, the efficient allocation of resources. Real option analysis can help maximise value by guiding project design and phasing. Modelling the precise array of options available on any investment can be a computationally complex task. This topic, however, is general to investment appraisal across most sectors of the economy, with no particular remarks to make about aviation. Suffice it to say that the valuation of timing and phasing is very much project-specific, and the evaluation should be tailored to reflect project circumstances. The use of real option analysis on a level beyond a simple, first, rough estimate, almost inevitably requires detailed work on the timing aspects specific to the project being appraised.

Strategic interaction between competitors can also play an important role in project appraisal. This would consist of the project promoter building alternative scenarios about how competitors may be expected to react to alternative investment strategies. The investment decision therefore becomes contingent on expected competitor reactions. This is important in particular for the aeronautical industry and for airports, both sectors operating in competitive markets characterised by product differentiation and sunk costs. Where there is more limited scope for product differentiation and sunk costs are few, as in many airline market segments, the role for strategic interaction models is more limited. This is because the investor can be expected to face competition from a virtually endless series of competitors, all essentially behaving similarly.

For ATM, where there is little scope for competition, the role that strategic interaction plays in the investment decision is naturally marginal. Competitive interaction calls for managers to appraise a wider range of scenarios, each depending on competitor response. In that sense, rather than adding detail to the models, the investment appraisal exercise is enriched to explore a wider array of circumstances. Ideally, such analysis would make use of insights offered by both industrial organisation and game theory into the incentive profiles of the various competitors, and their likely responses.

Possible refinements are many although, as has been mentioned above, investment appraisal is always about constructing models, and models are imperfect representations of reality. Any analysis can be refined by adding variables and by increasing the level of detail with which such variables are treated. But in investment appraisal, as in many other activities, diminishing returns soon set in. It is up to the analyst to judge whether the extra effort required in adding complexity to the analysis pays off in terms of new insights or enhanced estimate reliability. That is, whether it is likely to make a difference to the investment decision making. It would be ironic if in carrying out an economic appraisal aimed at attaining an efficient allocation of resources, the analyst were to end up inefficiently allocating too many resources to the appraisal. When making such a judgement, the analyst should bear in mind that the investment appraisal involves making assumptions about future conditions, assumptions that become stronger as the projections reach further into the project life. There is little point in devoting many resources to adding detail about conditions observable in year 1, when the following 19 years of the estimated project life (itself often an expectation or a convention) are increasingly uncertain, so that each detail added must then rely on new suppositions. The intended message is that economic appraisals need not be cumbersome or expensive exercises. Often, a small number of key variables will prove sufficient to build a fairly reliable picture about the merits of an investment project.

Appraisal resources could then perhaps be more productively deployed to assist the project conception decision making process.

This suggestion points towards the underlying rationale for conducting economic appraisals. Aviation uses large amounts of resources, and whereas it generates much value, it is not free from waste or from large potential losses. Managers, regulators and planners need to make informed choices regarding the conception of the project, including also whether to carry it out at all. When making such choices, an economic evaluation of the investment identifies areas of risk and opportunity that escape a financial analysis. More generally, conducting an economic appraisal gives as comprehensive a view as can be gathered about the intrinsic viability of an investment, both to society and to the investor, whether from the public or private sector.

References

Averch, H. and Johnson, L.L. 1962. Behaviour of the firm under regulatory constraint, *American Economic Review*, 52(5), 1052–1069.

Banister, D. and Berechman, J. 2001. *Transport Investment and Economic Development*. London: UCL Press.

Belenky, P. 2011. *The Value of Travel Time Savings: Departmental Guidance for Conducting Economic Evaluations*. Revision 2. Washington DC: U.S. Department of Transportation. [Online.] Available at: http://www.dot.gov/sites/dot.dev/files/docs/vot_guidance_092811c.pdf (accessed: 16 September 2013).

Belleflamme, P. and Peitz, M. 2010. *Industrial Organization: Markets and Strategies*, Cambridge, UK: Cambridge University Press.

Boardman, A., Greenberg, D., Vining, A. and Weimer, D. 2014 *Cost-Benefit Analysis: Concepts and Practice*. 4th Edition. Harlow, UK: Pearson.

Brealey, R.A., Myers, S.C. and Allen, F. 2008. *Principles of Corporate Finance*. 9th Edition. New York: McGraw-Hill Irwin.

Brent, R.J. 1998. *Cost-Benefit Analysis for Developing Countries*. Cheltenham, UK: Edward Elgar.

Campbell, C. and Brown, R. 2003. *Benefit-Cost Analysis: Financial and Economic Appraisal Using Spreadsheets*. Cambridge, UK: Cambridge University Press.

CE Delft 2002. *External Costs of Aviation*. Report conducted by CE Solutions for environment, economy, and technology, Delft, commissioned by Umweltbundesamt [Online.] Available at: http://www.ce.nl/art/uploads/file/02_7700_04.pdf (accessed: 9 July 2013).

Clark, P. 2007. *Buying the Big Jets: Fleet Planning for Airlines*. 2nd Edition. Aldershot, UK: Ashgate.

Coldren, G.M., Koppelman, F.S., Kasturirangan, K. and Mukherjee, A. 2003. *Air Travel Itinerary Share Prediction: Logit Model Development at a Major U.S. Airline*. Conference Proceedings of the Transportation Research Board 82nd Annual Meeting. Washington DC: Transportation Research Board.

Crompton, J.L. 2006. Economic impact studies: instruments for political shenanigans? *Journal of Travel Research*, 45 (August), 67–82.

Daley, B. 2010. *Air Transport and the Environment*. Farnham, UK: Ashgate.

de Neufville, R. and Odini, A. 2003. *Airport Systems: Planning, Design, and Management*. New York: McGraw-Hill.

de Rus, G. 2010. *Introduction to Cost Benefit Analysis: Looking for Reasonable Shortcuts*. Cheltenham, UK: Edward Elgar.

Dixit, R.K. and Pindyck, R.S. 1994. *Investment Under Uncertainty*. Newark, NJ: Princeton University Press.

Doganis, R. 2010. *Flying Off Course: Airline Economics and Marketing*. 4th Edition. Abingdon, UK: Routledge.

Douglas, G.W. and Miller III, J.C. 1974. Quality competition, industry equilibrium, and efficiency in the price-constrained airline market, *American Economic Review*, 64, 657–69.

Eurocontrol 2009. *Standard Inputs for Eurocontrol Cost Benefit Analyses*. Edition 4.0, October. [Online.] Available at: www.eurocontrol.int (accessed: 31 July 2013).

European Commission 2008. *Guide to Cost-Benefit Analysis of Investment Projects*. Brussels: Directorate General Regional Policy. [Online]. Available at: http://ec.europa.eu/regional_policy/sources/docgener/guides/cost/guide2008_en.pdf . [Online.] (accessed: 31 July 2013).

Federal Aviation Administration 1999. *FAA Airport Benefit-Cost Analysis Guidance*. FAA Office of Aviation Policy and Plans, United States Department of Transport. [Online.] Available at: http://www.faa.gov/regulations_policies/policy_guidance/benefit_cost/ (Accessed 31 August 2013).

Federal Aviation Administration 2010. *Addendum to FAA Airport Benefit-Cost Analysis Guidance*. FAA Office of Policy and Plans, United States Department of Transport. [Online.] Available at: http://www.faa.gov/regulations_policies/policy_guidance/benefit_cost/ (Accessed: 31 August 2013).

Fishwick, F. 1993. *Making Sense of Competition Policy*. London: Kogan Page.

Garrow, L.A. 2010. *Discrete Choice Modelling and Air Travel Demand: Theory and Applications*. Burlington, VT: Ashgate.

Ghobrial, A. 1993. A model to estimate the demand between U.S. and foreign gateways. *International Journal of Transport Economics*, 20(3), 271–83.

HEATCO 2006. *Developing Harmonised European Approaches for Transport Costing and Project Assessment. Deliverable 5: Proposal for Harmonised Guidelines*. Report by the Transport Research & Innovation Portal (TRIP), funded by the European Commission 6th Framework Programme for Research and Technological Development. [Online.] Available at: http://www.transport-research.info/web/projects/project_details.cfm?id=11056 (accessed: 9 July 2013).

Hess, S., Adler, T. and Polak, J.W. 2007. Modelling airport and airline choice behaviour with the use of stated preference survey data. *Transportation Research*, Part E, 43(3), 221–33.

Hummels, D. 2001. *Time as a Trade Barrier*. West Lafayette, IN: Purdue University. [Online.] Available at: http://www.krannert.purdue.edu/faculty/hummelsd/research/time3b.pdf (accessed: 28 July 2013).

Hummels, D. and Nathan Associates Inc. 2007. *Calculating Tariff Equivalents for Time and Trade*. Washington DC: United States Agency for International Development (USAID). [Online.] Available at: http://www.krannert.purdue.edu/faculty/hummelsd/research/tariff_equivalents.pdf.(accessed: 31 July 2013).

Johnson, C. and Briscoe, S. 1995. *Measuring the Economy: A Guide to Understanding Official Statistics*. New Edition. London: Penguin.

Jorge, J.D. and de Rus, G. 2004. Cost-benefit analysis of investments in airport infrastructure: a practical approach. *Journal of Air Transport Management*, 10, 311–26.

Kay, J. 1995. *Foundations of Corporate Success*. Oxford: Oxford University Press.

Kodukula, P. and Papudesu, C. 2006. *Project Valuation Using Real Options: A Practitioner's Guide*. Fort Lauderdale, FL: J Rosh.

Koller, T., Goedhart, M. and Wessels, D. 2010. *Valuation: Measuring and Managing the Value of Companies*. 5th Edition. Hoboken, NJ: Wiley.

Lambin, J.J., Chumpitaz, R. and Schuiling, I. 2007. *Market-Driven Management: Strategic and Operational Marketing*. 2nd Edition. New York: Palgrave Macmillan.

Ling, F.-I., Lin, K. and Lu, J.-L. 2005. Difference in service quality of cross-strait airlines and its effect on passengers' preferences. *Journal of the Eastern Asia Society for Transportation Studies*, 6, 798–813.

Londero, E.H. 2003. *Shadow Prices for Project Appraisal: Theory and Practice*. Cheltenham, UK: Edward Edgar.

Lu, J.L. and Tsai, L.N. 2004. Modelling the effect of enlarged seating room on passenger preferences of domestic airlines in Taiwan. *Journal of Air Transportation*, 9(2), 83–96.

Martin, S. 2010. *Industrial Organization in Context*. Oxford: Oxford University Press.

Moss, D.A. 2007. *A Concise Guide to Macroeconomics: What Managers, Executives, and Students Need to Know*. Boston, MA: Harvard Business School Press.

Motta, M. 2004. *Competition Policy: Theory and Practice*. Cambridge, UK: Cambridge University Press.

Nordas, H.K. 2006. Time as a trade barrier: implications for low-income countries. *OECD Economic Studies*, 42(1), 137–67. [Online.] Available at: http://www.oecd.org/dataoecd/29/26/38698148.pdf (accessed: 31 July 2013).

Smit, H.T.J. and Trigeorgis, L. 2004. *Strategic Investment: Real Options and Games*. Princeton, NJ: Princeton University Press.

Trigeorgis, L. 1996. *Real Options: Managerial Flexibility and Strategy in Resource Allocation*. Cambridge, MA: The MIT Press.

Varian, H.R. 1992 *Microeconomic Analysis*. 3rd Edition. London, UK: W.W. Norton & Co.

Vasigh, B., Fleming, K. and Tacker, T. 2008. *Introduction to Air Transport Economics: From Theory to Applications*. Farnham, UK: Ashgate.

Vose, D. 2008. *Risk Analysis: A Quantitative Guide*. 3rd Edition. Chichester, UK: Wiley.

Winston, C. and de Rus, G. (eds) 2008. *Aviation Infrastructure Performance: A Study in Comparative Political Economy*. Washington DC: Brookings Institution Press.

World Bank 2005. *Treatment of Induced Traffic*. Transport Note TRN-11. Washington DC: World Bank:. [Online.] Available at: http://siteresources.worldbank.org/INTTRANSPORT/Resources/336291-1227561426235/5611053-1231943010251/trn-11EENote2.pdf. (accessed: 31 July 2013).

Index

aeronautical industry
    components 55, 191–3, 209–22
    economics xiv, xvi, 31–2, 55, 193–4, 225
    final assembler 191, 193–4, 195–209
    identified 191–2
    original equipment manufacturer 191
air navigation service provider 46, 55, 147–58, 223; *see also* air traffic management; flight
air traffic management
    and airline behavior 147–8, 155, 223
    capacity 141–2, 148
    charges, *see* charge, air navigation
    congestion, *see main entry*
    economics xiv, 46–8, 55, 142–3, 192
aircraft; *see also* willingness to pay, aircraft
    landing charge, *see* charge
    manufacturing, *see* aeronautical industry
    noise, *see* externality
    take-off weight 41, 115, 126, 149
    option on, *see* real option
    size 41, 47–50, 115, 117, 126–38, 141–52, 155
airline
    environmental impact, *see* externality
    economics xiv, 31–2, 55, 120–22, 127, 206–7, 225
    fleet planning 48, 126–8, 143–8, 159–60, 167–8, 192, 194, 224
    load factor 16–17, 127–8
    scheduling 16–17, 46, 48, 98, 115, 121, 127, 131–2, 142, 148, 155, 161
airport
    airside, *see main entry*
    capacity, *see* runway; terminal

catchment area 55, 59, 70, 91, 110
charges, *see* charge
congestion, *see main entry*
economics xiv, 6, 55, 91; *see also* economies (cost)
greenfield 70–93, 98–9
runway, *see main entry*
slots 43, 48, 98, 126–7
terminal, *see main entry*
airside 41, 47–50, 69, 74, 99–100, 139, 141; *see also* runway
ANSP, *see* air navigation service provider
ATM, *see* air traffic management
Averch-Johnson effect 102, 122, 139; *see also* regulation; overinvestment

Black-Scholes, *see* real option

cargo, *see* freight
cash flow, *see* financial appraisal
charge, *see also* non-aeronautical revenues; price; yield
    air navigation 149–54, 157
    aeronautical 6–7, 84–91, 108, 118–25, 138
    approach 158
    landing 24, 117, 120, 127, 158
    regulation, *see main entry*
comfort (and discomfort) 1, 15, 27–9, 48, 224
competition
    high (but less than perfect) 31, 41, 159
    inter-modal xvi, 47, 162, 179–89
    monopolistic xii, xiv, 52, 55–6, 102, 193, 201–6, 209
    oligopolistic xiv, 52–5, 194–9, 201, 209
    perfect 31, 35, 46, 52–5, 179, 196, 209

competitive advantage xi, xiii, xv, 2, 7–8, 46–7, 84, 110, 159, 178, 189, 206, 210; *see also* price, power; producer surplus
congestion 1, 26, 32, 43–4, 60, 81–2, 92–3, 109, 127, 223
consumer surplus xv, 8, 11, 32, 62, 82–5, 93, 101–2, 108, 110, 120, 122, 178, 181, 184–8; *see also* competitive advantage; producer surplus
continuous climb departure 154
continuous descent approach 154
cost
    economies, *see main entry*
    generalised, *see main entry*
    sunk xiv, xvi, 32, 102, 192–4, 210, 225 *see also* indivisibilities; risk, and sunk costs
    switching, 6–7, 32–3, 53; *see also* competitive advantage
cost-benefit analysis (and economic appraisal); *see also* consumer surplus; discount rate; externality; generalised cost; real option, economic; subsidy; welfare
    and demand forecasting 161; *see also* delay
    employment 30, 39–40, 196
    European Commission, *see main entry*
    explained xii–xvi, 7–12, 51, 82, 191, 196, 226–7
    financial vs. economic appraisal 7–9; *see also* producer surplus
    multiplier effects (income or employment) 3, 29, 38
    secondary market(s) 26, 33–7, 43, 54–5, 62, 193, 196–7, 204–5, 207
    value of air transport 160, 184–9
    wider economic benefits xv, 2, 29–40

deadweight loss, *see* welfare
delay
    frequency 16–17, 27, 29, 47–50, 73–4, 128–33, 141, 161
    propagation 43, 81, 223

    schedule 16, 61, 161
    stochastic 16, 27, 128, 161
demand, *see* traffic
Department of Transportation (of the United States of America) 18
departure frequency, *see* flight, frequency
discomfort, *see* comfort
discount rate
    and risk and uncertainty *see* risk, and discount rate
    social 9–11, 220

economic appraisal, *see* cost-benefit analysis
economic return, *see* cost-benefit analysis, financial vs. economic appraisal
economies (cost); *see also* indivisibilities
    density 52, 93, 100, 139, 224, 225
    scale (and returns to) 26, 32, 34–6, 49, 52, 62, 82, 93, 100, 132, 224, 225
    scope 52, 224, 225
environment, *see* externality
EMOSIA 224
employment, *see* cost-benefit analysis
Eurocontrol 149–50, 184, 224
European Commission
    and cost-benefit analysis 12, 18
    HEATCO 18–19, 25, 73
externality; *see also* delay, propagation; regulation, environmental
    air pollution (and air quality) 83, 127–8, 155, 157–8, 184, 197
    environmental cost xi, 83, 99, 133, 160, 178, 204–9
    greenhouse gas 6, 24–5, 83, 127, 142, 160, 204
    internalised 8, 24–5, 83, 99, 116, 142, 155, 160, 164, 196–7, 204–9
    noise 6, 24–5, 38, 74, 83, 117, 127, 142, 155–8, 180, 184–5, 197, 204–9
    in secondary market 54, 204–6; *see also main entry*

Federal Aviation Administration (of the United States of America) 43, 81, 223

Index 235

financial appraisal
   and economic appraisal xiii, 1, 4–9, 11, 12, 20, 26, 51–2, 81–2, 160, 162, 218
   and generalised cost 29
   discounted cash flow and real options analysis 173–5, 210–11, 217–18
flight
   efficiency 154–8; *see also* continuous climb departure; continuous descent approach
   frequency, *see* delay, frequency
freight (or cargo) 1, 13, 15, 17–19, 22, 32–3, 41, 46–7, 52, 57, 177, 191–2, 225; *see also* time, freight
frequency delay, *see* delay

GDP, *see* national income
generalised cost of transport (or travel)
   behavioural (or user) 26–9, 47, 71, 116–17, 180
   defined 26–9
   estimation 71–4, 84–5
   and land value 36–8
   and local economy 30, 33, 62, 179–83
   private, *see* behavioural *sub-entry*
   and project benefits 42–7, 58–62, 65–6, 98–9, 117, 129–30, 184–5
   and subsidy, *see main entry*
   total (or economic) 26, 28, 180
   user, *see* behavioural *sub-entry*
greenhouse gas, *see* externality

indivisibilities 6, 32, 50, 221
International Air Transport Association (IATA) 74
International Civil Aviation Organisation (ICAO) 116, 149, 153

land price (and value) 36–8, 74
landside 41–7

market, *see* competition; secondary market
multiplier effects, *see* cost-benefit analysis

national income (and GDP); *see also* generalised cost of transport, and local economy
   and demand for air transport 70–71; *see also* value of time *sub-entry*
   identified 178–9
   multiplier, *see* cost-benefit analysis, multiplier effects
   and runways 133, 138
   and the value of aviation 160–61, 178–83, 224
   and value of time 18, 80, 91, 133
   and welfare 30–31, 38, 179
noise, *see* externality
non-aeronautical revenues 75, 84–5, 91

oligopoly, *see* competition
option, *see* real option
original equipment manufacturer, *see* aeronautical industry
overinvestment 69, 101–14, 139, 149, 153; *see also* Averch-Johnson effect; regulation

pollution, *see* externality
producer surplus 36, 62–7, 82, 85, 99, 109, 138, 148, 152, 178, 188; *see also* competitive advantage; consumer surplus
price; *see also* regulation, price
   land, *see main entry*
   power xii, xiii, 6, 8, 55, 110, 127, 162, 195; *see also* competitive advantage
   shadow (and shadow wage) 20–21, 25, 28–9, 39–40, 195–6

quality 41, 46, 48, 53, 58, 81, 103, 115, 132, 153, 201; *see also* airport; comfort

R&D, *see* research, development and innovation
rate of return, see cost-benefit analysis, financial vs. economic appraisal
real option; *see also* financial appraisal, discounted cash flow and real options analysis

on aircraft 168–77
backward induction 214
binomial tree (or binomial lattice) 212–18, 220–22
Black-Scholes 171, 174–5, 211, 216
economic 218–22
exercise (or strike) price, 173, 175, 212, 215, 217, 221
expiry 174, 211
terminal node (or cell) 214
value (interpretation) 168–72, 209–12 see also Black-Scholes; binomial tree sub-entries
regulation; see also Averch-Johnson effect; overinvestment
and economic appraisal xii, 5, 9, 13, 25
environmental 161, 210, 214, 222
of investment projects 110, 122, 222
price (or return, or monopoly) 6, 100, 102, 108, 120–21, 139, 149–54
regulatory asset base 4, 122, 153, 157–8
safety 22–3; see also main entry
research, development and innovation xiv, xvi, 6, 191–2, 209–22
risk (and uncertainty); see also real option
accident, see safety
analysis xiv, 11–12
and competition 168, 211
and discount rate 10–12, 31, 100–101, 110, 210–11, 217–18, 219
identified by economic appraisal xiii, xv, 2, 8–9, 25, 167, 227; see also tax, risk
and sunk costs 193–4
rule of a half 45, 81, 203; see also welfare, deadweight loss
runway
capacity 50, 115, 132–3
enlargement 115–25, 157
new 47, 126–39

safety (and accident risk) 1, 3, 15, 22–3, 29, 73, 80, 83, 109, 153, 180
secondary market 26, 29–31, 33–6, 43, 54–5, 62, 188, 193, 196–7, 204–7;

see also externality, in secondary market; land price
shadow price, see price
schedule delay, see delay
social discount rate, see discount rate
stochastic delay, see delay
subsidy; see also tax
to aeronautics 195, 205
to airport 56, 91, 100
and economic appraisal 9, 160, 178, 181
and generalised cost of transport 26, 28
to high-speed rail 185–8
and price 5, 21, 167
(assistance to finance) research 221–2

tax; see also subsidy
as distortion and transfer 5, 20, 80, 82, 181, 195, 203
on externality 8, 90, 158, 163–4, 197
and overinvestment 108–10
rebates 91
risk of 8; see also risk, identified by economic appraisal
treatment (in the book) 13
terminal
expansion 57, 69, 74, 92–9, 101–14
level of service 41, 74
node see real option, terminal node
time; see also delay
access and egress 15, 16, 72–3, 179
departure or arrival (or flight) 16, 27, 43, 46–7, 59–61, 98–9, 128–31, 142, 155; see also frequency delay; traffic, diverted
freight 17, 19, 47
in-vehicle 47, 142, 180
processing (and congestion) 16, 43, 72, 81, 179–80, 225
savings 1, 4, 36, 59, 116, 181, 189
as switching (or diversion) cost 6–7, 188
travel (door to door) 16–17, 27–9, 142, 154, 180
value (or willingness to pay for saving) 17–19, 33, 36, 37, 47, 73, 80, 91, 116, 132–3, 143, 180, 185, 224–5

Index

traffic
    deterred 44–7, 60–63, 71–81, 85–90, 92–3, 103, 109, 129–30, 152; *see also* generated *sub-entry*
    diverted 44–5, 58–66, 73, 80–84, 90, 93, 98–9, 101–9, 117–20, 128–30, 132, 147–8, 152, 185
    existing 44–45, 108, 170
    generated 8, 44–5, 58–66, 71–4, 81–5, 90, 99, 101–9, 117–21, 127, 129–30, 148, 152, 157, 185, 188–9, 205; *see also* deterred *sub-entry*
    growth 50, 62, 92, 110, 129, 148, 159

uncertainty, *see* risk (and uncertainty)

WACC, *see* discount rate
welfare; *see also* cost-benefit analysis
    and income (or output) 5, 24, 30–31, 35, 38
    deadweight loss 45, 67, 109, 202–3, 205; *see also* rule of a half
    transfer 62, 67, 90–91, 108, 152, 201–3

willingness to pay
    and (or for) aircraft 196, 224
    for comfort 15, 27–8
    and economic appraisal 7, 30, 32–3, 45, 184, 188
    property 36
    for safety 73
    for time, *see* time, value

yield
    aeronautical 84–5
    non-aeronautical 84, 91
    passenger 84–5, 93, 98, 117, 164, 170–71